T0139183

Wireless Sensor Multimedia Networks

Architectures, Protocols, and Applications

Wireless Sensor Multimedia Networks

Architectures, Protocols, and Applications

Edited by
Mohamed Mostafa A. Azim • Xiaohong Jiang

CRC Press
Taylor & Francis Group
Boca Raton London New York

CRC Press is an imprint of the
Taylor & Francis Group, an **informa** business

CRC Press
Taylor & Francis Group
6000 Broken Sound Parkway NW, Suite 300
Boca Raton, FL 33487-2742

Printed on acid-free paper
Version Date: 20150624

International Standard Book Number-13: 978-1-4822-5311-5 (Hardback)

Library of Congress Cataloging-in-Publication Data

Wireless sensor multimedia networks : architectures, protocols, and applications / editors, Mohamed Mostafa A. Azim and Xiaohong Jiang.
pages cm
Includes bibliographical references and index.
ISBN 978-1-4822-5311-5 (alk. paper)
1. Wireless sensor networks. 2. Multimedia communications. I. Azim, Mohamed Mostafa A. II. Jiang Xiaohong.

TK7872.D48W56667 2015
681'.2--dc23 2015021738

Visit the Taylor & Francis Web site at
http://www.taylorandfrancis.com

and the CRC Press Web site at
http://www.crcpress.com

Contents

Editors

Mohamed Mostafa A. Azim earned his BSc in electrical engineering from Cairo University, Giza, Egypt, in 1994; his MSc jointly from the Fontys University, Eindhoven, the Netherlands, and the Eindhoven University of Technology, Eindhoven, the Netherlands, in 1997; and his PhD in computer sciences from Tohoku University, Sendai, Japan, in 2006. He is an associate professor of electronics technology, Beni-suef University, Beni-suef, Egypt. In 2008, he joined the College of Computer Science and Engineering at Taibah University, Tayba, Kingdom of Saudi Arabia. In 2012, he became the chair of the Networks and Communication Systems Department. Dr. Azim has authored the book, *Optical Networks: A Restoration Perspective with Active Restoration*. He has also authored several research papers in the most reputed journals and presented them at conferences. He is a working group member of the IEEE 1903 standard on Next-Generation Service Overlay Networks (NGSON). He is on the editorial boards of the *International Journal of Sensor and Related Networks* and *International Journal of Communication Networks and Information Security* (IJCNIS). His current research interests include network modeling and simulation, routing and security protocols for wireless sensor networks and wireless multimedia sensor networks, and optical networks survivability.

Xiaohong Jiang earned his BS, MS, and PhD in 1989, 1992, and 1999, respectively, from Xidian University, Xi'an, China. He is currently a full professor at Future University, Hakodate, Japan. Before joining Future University, Dr. Jiang was an associate professor at Tohoku University from February 2005 to March 2010. He was an assistant professor at the Graduate School of Information Science, Japan Advanced Institute of Science and Technology (JAIST), from October 2001 to January 2005. Dr. Jiang was a JSPS (Japan Society for the Promotion of Science) postdoctoral research fellow at JAIST from October 1999 to October 2001. He was a research associate in the Department of Electronics and Electrical Engineering, University of Edinburgh, from March 1999 to October 1999. Dr. Jiang's research interests include computer communications networks, mainly wireless networks and optical networks; interconnection networks for massive parallel computing systems; routers/switches design for high-performance networks; and network security. He has published over 230 technical papers in premium international journals and presented at conferences, including over 30 papers published in top IEEE journals and top IEEE conferences, such as *IEEE/ACM Transactions on Networking*, *IEEE Journal of Selected Areas on Communications*, and IEEE INFOCOM. Dr. Jiang was the winner of both the Best Paper Award and the Outstanding Paper Award of IEEE WCNC 2012, IEEE WCNC 2008, IEEE ICC 2005-Optical Networking Symposium, and IEEE/IEICE HPSR 2002. He is a senior member of IEEE and also a member of the Institute of Electronics, Information and Communication Engineers.

Contributors

Rania Ahmed Abul-Seoud earned her BSc in communications and electronics engineering from Fayoum University, Fayoum, Egypt, in 1998, her MS in artificial intelligence from the Faculty of Computer Engineering at Fayoum University in 2005, and her PhD in biomedical engineering, artificial intelligence, and its applications to biomedical informatics from Fayoum University in 2008. She is the author of several research papers published in highly reputable journals and conference proceedings. Her current research interests are artificial intelligence and its different applications, biomedical informatics, and routing and security protocols for wireless sensor networks.

Ozgur B. Akan earned his PhD in electrical and computer engineering from the Broadband and Wireless Networking Laboratory, School of Electrical and Computer Engineering, Georgia Institute of Technology, Atlanta, Georgia, in 2004. He is currently a full professor with the Department of Electrical and Electronics Engineering, Koc University, Istanbul, Turkey, and the director of the Next-Generation and Wireless Communications Laboratory. His current research interests are wireless communications, nanoscale and molecular communications, and information theory. He is an associate editor of the *IEEE Transactions on Communications*, the *IEEE Transactions on Vehicular Technology*, the *International Journal of Communication Systems* (Wiley), the *Nano Communication Networks* journal (Elsevier), and the *European Transactions on Technology*.

Muhammad Mahbub Alam earned his BS in applied physics and electronics and his MS in computer science in 1998 and 2000, respectively, from the University of Dhaka, Dhaka, Bangladesh. He earned his PhD from the Department of Computer Engineering, Kyung Hee University, South Korea, in 2008. Currently, Dr. Alam is working as a professor in the Department of Computer Science and Engineering at the Islamic University of Technology, Gazipur, Bangladesh. His research interests include wireless and mobile networking and performance modeling and analysis of networking systems.

Raghied Mohammed Atta earned his BSc in electronics and communications engineering from the Faculty of Engineering, Cairo University, Cairo, Egypt, his MSc in signal processing from the Faculty of Electronic Engineering, Menoufia University, Menouf, Egypt, in 1992, and his PhD from the Department of Engineering, University of Cambridge, Cambridge, UK. Currently, he is an associate professor in the Electrical Engineering Department, Engineering College, Taibah University, Madinah, Kingdom of Saudi Arabia.

Guilin Chen earned his BS in mathematics from Anhui Normal University, Wuhu, China, in 1985, and MS in computer applications from Hefei University of Technology, Hefei, China, in 2007. Dr. Chen is currently a professor in the School of Computer and Information Engineering at Chuzhou University, Chuzhou, China. His main research interests include cloud computing, the Internet of Things, and big data.

Md. Abdul Hamid earned his BE in computer and information engineering in 2001 from the International Islamic University Malaysia (IIUM), Seoul, Yongin, South Korea. In 2002, he became a lecturer at the Computer Science and Engineering Department, Asian University of Bangladesh, Dhaka, Bangladesh. He earned his PhD from the Computer Engineering Department at Kyung Hee University, Suwon, South Korea, in August 2009. In September 2009, he became an assistant professor in the Department of Information and Communications Engineering at Hankuk University of Foreign Studies, Yongin, South Korea. He then joined Green University of Bangladesh and worked as an assistant professor in the Department of Computer Science and Engineering from September 2012 to May 2013. Dr. Hamid is currently serving as a faculty member in the Department of Computer Engineering, Taibah University, Madinah, Kingdom of Saudi Arabia. His research interests include wireless sensor, mesh, ad hoc, and opportunistic networks with particular emphasis on network security, reliability, fairness, and quality-of-service issues.

Abu Raihan Mostofa Kamal is currently serving as a faculty member in the Department of Computer Science and Engineering at the Islamic University of Technology, Gazipur, Bangladesh. He worked in the area of embedded networked systems, specifically wireless sensor networks, at the School of Computer Science and Informatics, University College Dublin, Dublin, Ireland. His PhD work focused on enhanced reliability in wireless sensor networks. He completed his PhD research under the supervision of Dr. Chris Bleakley in 2013. The core area of Dr. Kamal's research includes data and network fault detection in sensor networks. He completed his MSc in information and communication security at the Royal Institute of Technology, Stockholm, Sweden, in 2004. Dr. Kamal also worked as a postdoctoral researcher at the Nimbus Centre, Cork Institute of Technology, Cork, Ireland, between May 2013 and May 2014.

Nour El-Deen Mahmoud Khalifa earned his MS and PhD in 2009 and 2013, respectively, both from the Faculty of Computers and Information, Information Technology Department, Cairo University, Cairo, Egypt. Currently, he is a research doctor at the Faculty of Computers and Information at Cairo University. Dr. Khalifa's academic and research specialties are information technology, multimedia, wireless sensor networks, communication protocols in WMSNs, cryptography, wireless communication security, network security, multimedia security, and NS2.

Xiaolan Liu earned her MS in information and communication engineering from Dalian Maritime University, Dalian, China, in 2009. Since 2009, she has been with the School of Computer and Information Engineering at Chuzhou University, Chuzhou, China. Her research interests include wireless camera sensor networks and barrier coverage.

Mustafa Ozger earned his BSc in electrical and electronics engineering from Middle East Technical University, Ankara, Turkey, in 2011, and MSc in electrical and electronics engineering from Koc University, Istanbul, Turkey, in 2013. He is currently a research assistant in the Next-Generation and Wireless Communication Laboratory and pursuing his PhD at the Electrical and Electronics Engineering Department, Koc University, Istanbul, Turkey. His current research interests include cognitive radio networks and cognitive radio sensor networks.

Ecehan B. Pehlivanoglu earned his BSc in electrical and electronics engineering from Middle East Technical University, Ankara, Turkey, in 2011. He is currently a research assistant at the Next-Generation Wireless and Communication Laboratory while pursuing his PhD at the Electrical and Electronics

Engineering Department, Koc University, Istanbul, Turkey. His research interests include cognitive radio, cognitive radio sensor networks, and nanoscale communications.

Ahmed Hussein Abbas Salem earned his BSc in communications and electronics engineering from Fayoum University, Fayoum, Egypt, in 2014. He is the author of several research papers published in a number of conference proceedings. His current research interests are network communications and routing, security protocols, and applications for wireless sensor networks.

Pinar Sarisaray-Boluk earned her BS and MS in computer engineering from Karadeniz Technical University, Trabzon, Turkey. She earned her PhD in the Computer Engineering Department at Istanbul Technical University, Istanbul, Turkey. In 2013, she did postdoctoral training at the Computer Science Department of Southern Illinois University, Carbondale, Illinois. Currently, she is serving as an assistant professor in the Software Engineering Department at Bahcesehir University, Istanbul, Turkey. Her research interests are wireless sensor networks, network security, and software development, analysis, and design.

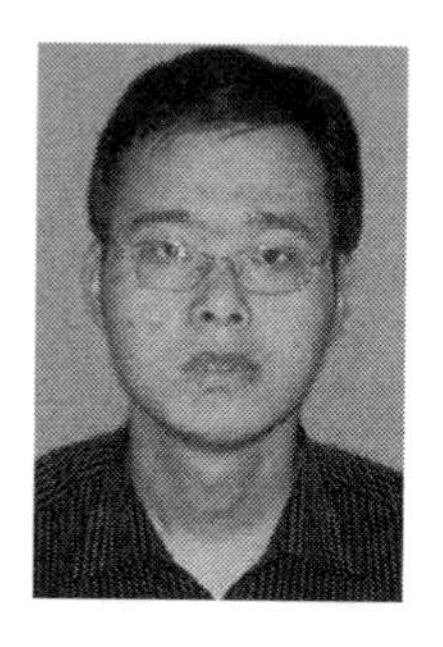

Bin Yang earned his BS and MS in computer science from Shihezi University, Shihezi, China, in 2004, and from the China University of Petroleum, Beijing, China, in 2007, respectively. He is currently a PhD candidate at the School of Systems Information Science, Future University Hakodate, Hakodate, Japan, and is also a faculty member at the School of Computer and Information Engineering, Chuzhou University, Chuzhou, China. His research interests include performance modeling and evaluation, stochastic optimization, and control in wireless networks, LTE-A, and 5G networks.

Introduction

Wireless Sensor Multimedia Networks

Wireless sensor networks (WSNs) are a special class of ad hoc networks in which network nodes are tiny sensors with limited processing power, memory, and battery power. These sensors cooperate to convey messages from each node to the sink node or gateway node.

Wireless sensor multimedia networks (WSMNs) are a special category of WSNs in which sensor nodes are small cameras and microphones, as shown in Figure I.1.

WSMNs differ greatly from WSNs. In WSNs, network nodes cooperate to convey scalar data such as temperature, pressure, humidity, and light to the sink node. In WSMNs, the data being sent are multimedia data such as voice, image, or video. However, scalar data can also be handled by a WSMN node, by adding the required sensors.

WSMNs have appeared as a result of intensive research in various areas such as VLSI, MEMS, digital signal processing, communications, and networks. The availability of inexpensive CMOS cameras and microphones accelerated the use of WSMNs in the market. Currently, WSMNs are attracting a great amount of attention from academia and industry due to the variety of applications where they can be deployed. Moreover, WSMNs have many challenges in their design and deployment.

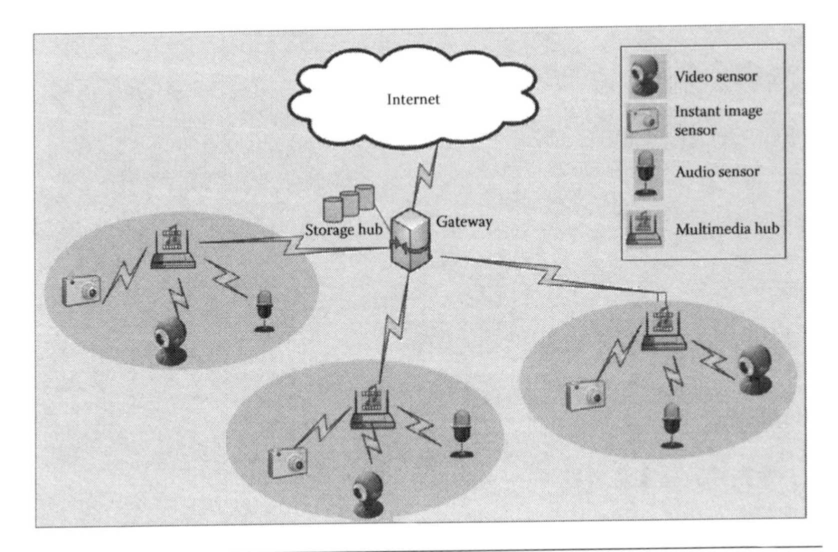

Figure I.1 WMSN architecture.

Benefits of WSMNs

WSMNs have the ability and flexibility to fuse and store multimedia content originating from different camera sources. Deploying multiple visual sensors as cameras has several benefits despite increasing the coverage and enlarging the field of view (FOV), but it also increases redundancy and reliability, as well. WSMNs offer the following benefits:

- Better enhanced FOV: This can be accomplished by using several cameras. When the FOV is dark, using a combination of cameras is beneficial to show the infrared and visible spectrum in the targeted scene.
- Larger FOV: Using several cameras enables a larger FOV. The main idea is to use several low-resolution cameras to trigger the few high-resolution cameras. These advanced cameras can then target the required event, using their pan–tilt–zoom functionality. This technique will provide the necessary quality at a lower cost.
- Adding several points of view: When an area such as a street needs to be monitored, one camera will not suffice, so several cameras are added to provide flexibility and several viewpoints.

These benefits have led to deploying WSMNs in a variety of applications:

- Surveillance, monitoring, and object protection applications: WSMNs help greatly in monitoring streets, public areas, and, more importantly, museums and borders.
- Storing footage of unusual events: WSMNs are able to record accidents, robberies, and traffic violations.
- Traffic congestion avoidance: Traffic in large cities can be monitored, which will help greatly, especially during rush hours.
- Health-care applications: In addition to sensing scalar data relevant to the patient—such as blood pressure, ECG, and heart rate—remote monitoring could be achieved by using motion sensors in conjunction with video and audio sensors. Recording the behavior of elderly people could be useful for research in the medical and health-care areas.
- Environmental applications: WSMNs can be used to monitor the environment and help warn of dangers such as global warming. For example, the polar ice caps could be monitored to predict the effects of global warming on the Earth's water levels.
- Habitat monitoring: Monitoring animals in certain areas will help in understanding the habits of wild animals and animals that prefer quiet settings.
- Localization services: Processing captured images and video may lead to locating missing objects or children and wanted criminals.

One of the challenging problems in WSMNs is routing the data collected from several cameras, especially if high resolution is required. Although providing better quality for images and videos is necessary, it shortens the network lifetime because the batteries will drain quickly. Therefore, minimizing energy consumption and developing energy-efficient protocols have become a must in WSMN.

WSMN Structure

A WSMN may have a simple structure, such as one or more nodes communicating directly to the base station through single-hop communication, which is known as star topology. Such a topology will not succeed in sizable applications. Therefore, it is essential to dissect WSMN operation. By dissecting the operation and the nodes to levels of hierarchy, energy consumption can be decreased and the network will be much more manageable and organized. WSMNs can have four main components:

1. Wireless multimedia nodes (WMNs): WMNs are the end points of the network. These nodes have a camera and a microphone. The WMN is mostly battery powered.
2. Wireless cluster head (WCH): The WCH receives data from several WMNs; it also removes scenes and events that are from the same FOV.
3. Wireless network node (WNN): A WNN acts as a relay node, delivering data from the network to the base station. Its main purpose is to decrease the distance between the network and the base station in order to decrease energy consumption.
4. The base station (BS): It collects the data on the PC for further processing and research applications.

The main goal of such a structure is decreasing energy consumption as much as possible.

The WSMN Node

The WSMN node is designed to be small in size with low power consumption and low cost. All WSMN nodes consist of the following four components:

1. The *communication unit* contains the transceiver, which is usually based on the ZigBee [1] (IEEE* 802.15.4) standard or 6LoWPAN (IPv6 over Low Power Wireless Personal Area Networks) [2]. These standards target low power consumption during transmission and reception processes. The simple difference between both standards is that 6LoWPAN supports relaying the sensor data to the Internet.
2. The *processing unit* will differ significantly from the processors used in normal WSNs, as the WSMN may contain a processor combined with a processor specific for image or

* IEEE, Institute of Electrical and Electronics Engineers.

video applications. Microcontrollers, DSPs, and application-specific processors are usually used in such nodes.

3. The *sensing unit* is the main difference between WSNs and WSMNs; WSMNs will have visual sensors like a camera and audio sensors such as a microphone.
4. The *power unit* is usually required to be portable, such as batteries.

WSMN Suppliers

In this section, we would like to point out one of the WSN suppliers, namely, Libelium [3]. Libelium is one of the high-end companies that succeeded in building a completely compatible, unique WSMN node. Libelium delivers a powerful, modular, easy-to-program open-source sensor platform for the Internet-of-Things enabling system. One of these products is the Waspmote. Waspmote is Libelium's advanced mote for WSNs; it can be integrated with different boards, such as the video camera board shown in Figure I.2.

Book Organization

Chapter 1: Multichannel, Multipath-Enabled, Quality-of-Service-Aware Routing for Wireless Multimedia Sensor Networks

This chapter presents a novel quality-of-service-aware routing protocol to support a high data rate in wireless multimedia sensor networks.

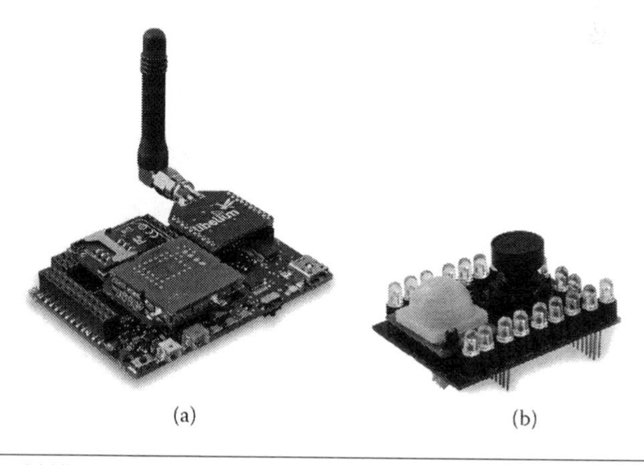

Figure I.2 (a) Libelium Waspmote and (b) a video camera board.

Chapter 2: Adaptation Techniques for Multimedia Communication in Wireless Sensor Networks

This chapter first surveys different factors affecting the design of communication protocols and multimedia application algorithms for efficient multimedia communication in sensor networks. It also presents different protocols and algorithms used in the transport, network, and MAC layers of WSMNs. It evaluates them in terms of multimedia transmission. Finally, some multimedia adaptation techniques are presented for each protocol to give an optimistic perspective for future deployment.

Chapter 3: Multimedia Communication in Cognitive Radio Ad Hoc and Sensor Networks

This chapter introduces the use of cognitive radio capability in sensor networks to increase the efficiency of overall spectrum utilization and to decrease the probability of collision and contention. This chapter also presents different factors influencing multimedia communications in cognitive radio ad hoc and sensor networks and demonstrates open research problems at different communication layers.

Chapter 4: Multimedia Streaming in Wireless Multimedia Sensor Networks

This chapter presents different multimedia streaming optimization techniques and highlights the differences between WSN and WSMN requirements for multimedia streaming.

Chapter 5: Coverage Problems for Wireless Multimedia Sensor Networks

This chapter describes the state of the art in influencing factors, deployment mechanisms, sensor selection, and performance metrics for WMSN node coverage. The authors present different types of coverage and state representative solution algorithms. Finally, they discuss existing problems and new research trends in some key realms.

Chapter 6: A Security Scheme for Video Streaming in Wireless Multimedia Sensor Networks

This chapter surveys different attacks on WSMNs. It also proposes a new security scheme that is appropriate for real-time video streaming.

The performance of the proposed scheme is verified by simulation. Open research problems in this area are also presented.

Chapter 7: Power Management for Wireless Multimedia Sensor Networks

This chapter presents several ways of optimizing the power consumption of existing WSMNs. It also identifies various unconventional sources of energy harvesting based on the available techniques and compares their advantages.

Mohamed Mostafa A. Azim
Cairo University, Giza

Xiaohong Jiang
Future University, Hakodate

References

1. IEEE 802.15 WPAN Task Group 4. http://www.ieee802.org/15/pub/TG4.html
2. Kushalnagar, N. et al. IPv6 over Low-Power Wireless Personal Area Networks (6LoWPANs): Overview, assumptions, problem statement, and goals, *IETF RFC 4919*, Aug 2007. http://www.rfceditor.org/rfc/pdfrfc/rfc4919.txt.pdf
3. Libelium. http://www.libelium.com/

1

Multichannel, Multipath-Enabled, Quality-of-Service-Aware Routing for Wireless Multimedia Sensor Networks

MD. ABDUL HAMID,
ABU RAIHAN MOSTOFA KAMAL,
AND MUHAMMAD MAHBUB ALAM

Contents

The majority of sensor network routing protocols consider energy efficiency as the main objective and assume data traffic with unconstrained delivery requirements. However, the introduction of image and video sensors demands a certain quality of service (QoS) from the routing protocols and underlying networks. Managing such real-time data requires both energy efficiency and QoS assurances to ensure efficient usage of sensor resources and accuracy of the collected information. In this chapter, we discuss this issue and present a novel QoS-aware routing protocol to support a high data rate for wireless multimedia sensor networks (WMSNs). With multichannel, multipath technology, routing decisions are made according to dynamic adjustment of the required bandwidth and path-length-based proportional delay differentiation (PPDD) for real-time data. To justify QoS requirements and to offer differentiated service, we classified and prioritized the sensor data. Finally, we evaluated the protocol performance through rigorous simulation under different scenarios. The simulation results demonstrate significant improvement in performance in terms of average end-to-end delay, average lifetime, network throughput, packet drop ratio (PDR), and delivery ratio. In particular, we delve into a performance evaluation of single-sink and multiple-sink schemes.

1.1 Introduction

Wireless sensor networks (WSNs) have unique characteristics, such as self-configuration, low cost, easy deployment, and distributed sensing capacity. As a result, the field of WSNs has enjoyed tremendous research attention over the past decade. Traditional WSNs are often deployed in hard-to-reach areas for prolonged durations to report various real-time scalar data such as temperature, humidity, and light intensity. Usually such networks are operated for a long time without any human intervention. Successful WSN deployment [1–3] has gained a new and promising dimension among the research community with the introduction of multimedia data such as image and video. As a result, applications can get high-dimensional data with increased accuracy both in event detection and periodic monitoring. Furthermore, for cost–benefit analysis, recent trends indicate that multiple applications with varied QoS requirements can be efficiently deployed in a single network [4,5]. Consequently, WMSNs have received a great deal of research attention in recent years. Traditional protocols for WSN cannot cope with MWSN gracefully because they were designed to handle scalar data from a single application. In this chapter, we present a novel QoS-aware packet delivery technique to support high data rates and delay bound requirements for WMSNs. The promising pace of technological growth has led to the design of sensor nodes with the capability of sensing the environment and producing multimedia data. However, because multimedia traffic contains images, video, audio, and scalar data, each merits a different metric. To accommodate a high data rate, designing an efficient routing protocol is of primary interest. The significance of such a protocol becomes clear with a few challenging and motivating facts. First, research challenges found in Ref. [6] state that existing data rates of about 40 and 250 kbit/s supported by the MICA2 and MICAz [7] motes are not geared to support multimedia traffic. Instead of improving the hardware and thus increasing cost, an alternate approach is to more efficiently utilize the available bandwidth. By using multiple channels in a spatially overlapped manner, the existing bandwidth can be leveraged to support multimedia applications. Second, the use of multipath technology has two clear advantages: (1) the load may be balanced so as not to overwhelm the limited buffers at the intermediate sensor nodes and (2) one path condition may not permit a high data rate for the entire duration of the event being monitored. By allowing multiple paths, the effective data rate of each path gets reduced and the application can be supported. This chapter presents a protocol that targets the application of WMSNs where sensors produce multimedia content from the deployed area to deal with both critical and general data. Applications may include critical condition monitoring and security surveillance tasks such as monitoring a volcano explosion, toxic gases, or a forest fire; military applications such as sniper or enemy detection; and civil applications such as the location of survivors for rescue services. Once a node detects an important event, fast and reliable delivery is required; late or failed delivery may cause disaster. In a real-time application such as multimedia streaming, delivered data can become useless in only a few milliseconds.

Though most currently deployed sensor networks use the same channel to communicate information among nodes, a significant number of current sensor node prototypes use radio modules capable of transmission on multiple channels. For example, the radio capabilities of the MICAz mote allow communication on multiple frequencies as specified in IEEE* standard 802.15.4. The idea of using multiple channels in wireless networks is not new. One study [8] has shown how the capacity of a static multichannel network scales as the number of nodes in the network increases. The authors show that it may be possible to build capacity optimal multichannel networks with as few as one interface per node. The authors of [9] present a multichannel defense mechanism against jamming attacks in WSNs by automatically and efficiently assigning nodes to different channels in the jammed area to defeat an attacker. The work presented in Ref. [10] introduces a control-theoretic approach for maximizing throughput in multichannel sensor networks by choosing node communication frequencies such that the total network throughput is maximized.

Classical multipath routing has been explored for two reasons. The first is load balancing (where traffic is split across multiple disjoint paths) and the second is to increase the likelihood of reliable data delivery (multiple copies of data are sent along different paths). Although a plethora of techniques have been developed for sensor networks, all protocols featured either multipath or multichannel technology. In fact, QoS provisioning is a challenging task for multimedia sensor networks because link capacity and delay vary continuously and may be bursty in nature [6]. Creating a QoS provisioning routing protocol with the efficient use of both multipath and multichannel technology to support the high data rate requirement for WMSNs has not been addressed. In this chapter, we present a mechanism for packet delivery over a multipath, multichannel-provisioned WMSN in which multimedia sensors ubiquitously retrieve multimedia contents from the environment. The initial version of this work can be found in Ref. [11]. Our main goal is to support a high data rate while maintaining the attainable delay so that packets can be delivered to the destination

* IEEE, Institute of Electrical and Electronics Engineers.

with their bandwidth and delay requirements. The main contributions of the chapter can be summarized as follows:

- We designed a QoS-aware routing protocol for WMSNs. More specifically, our design is based on multipath, multichannel technology, which influences how routing decisions for real-time and non-real-time multimedia traffic are made, using dynamic bandwidth adjustment and PPDD. To meet bandwidth requirements, the proposed technique provides network-wide dynamic bandwidth-adjustment options for the nodes in a distributed manner. To meet delay requirements, the proposed technique provides PPDD options, extending the idea of the PDD mechanism in Ref. [12].
- In order to define QoS in WMSNs, we classified data traffic, which facilitates the prioritization of different data packets.
- Rigorous simulation was carried out to evaluate the performance of the proposed design. The results show the advantages of our approach over the existing approach in terms of throughput, end-to-end delay, lifetime, PDR, and network-wise delivery ratio (NDR). Improvement was also noticed in multiple-sink scenarios, which is a natural solution for a very large network.

The rest of this chapter is organized as follows. Section 1.2 describes an overview of existing background works. Section 1.3 provides the network model and assumptions. Section 1.4 presents the proposed QoS-aware routing protocol in detail. Section 1.5 presents performance evaluation through simulation. Finally, Section 1.6 concludes this work with a summary and some future research challenges.

1.2 Background

A large number of studies have been carried out in this area since data collection became the most important aspect of WSNs. The findings of research challenges and the current status of the literature on multimedia communication in WSNs are presented in Refs. [6,13,14]. More specifically, factors influencing multimedia delivery over WSNs and currently proposed solutions for application, transport, and network layers are pointed out along with their shortcomings and open research issues. Cucchiara [15] gives a short overview of the hot topics in multimedia surveillance systems and introduces some research activities currently underway worldwide. For example, a multiflow real-time transport protocol described in Ref. [16] does not specifically address energy efficiency considerations in WMSNs, but is suited for real-time streaming of multimedia content by splitting packets over different flows. In Ref. [17], a wakeup scheme is proposed that tries to balance energy and delay constraints. In Ref. [18], the proposed protocol has an interesting feature: to establish multiple paths (optimal and suboptimal) with different energy metrics and assigned probabilities. Hence, it is inherently a multipath protocol with QoS measurements and a good fit for routing of multimedia streams in WSNs.

Recently, Hamid and Bashir [19] proposed a cross-layer QoS protocol for WMSNs. The protocol provides interaction between energy-based admission control, delay- and interference-aware routing, and dynamic duty cycle assignment in the MAC layer. Kim and Sung [20] proposed a scheme for efficiently and reliably delivering real-time multimedia streams in WSNs. To specify the property of streams, a multimedia stream is modeled as an (m,k)-firm stream that is known to have the characteristics of a weakly hard real-time system. A cross-layer framework is proposed in Ref. [21] to support QoS in WMSNs to enhance the number of video sources, given that the QoS constraint of each individual source is also preserved. Their goal is achieved by implementing Wyner–Ziv lossy distributed source coding at the sensor node with variable group of pictures size, exploiting multipath routing for real-time delivery and link adaptation to enhance the bandwidth under the given bit error rate. Touil et al. [22] analyzed energy consumption and evaluated the performance of the 802.11e enhanced distributed channel access (EDCA) with and without the contention-free burst (CFB) mechanism, compared with IEEE 802.11 DCF. They showed that the use of EDCA CFB gives better performance and offers a very good relationship between energy consumption and traffic performance, which is recommended in WMSNs.

A QoS provisioning multipath and multi-SPEED routing protocol (MMSPEED) was proposed [23]; MMSPEED spans the network layer and medium access control layer to provide QoS differentiation in timeliness and reliability. To support both best-effort and real-time traffic at the same time, a class-based queuing model was employed in Ref. [24]. The queuing model allows service sharing for real-time and non-real-time traffic. The bandwidth ratio r is defined as an initial value set by the gateway and represents the amount of bandwidth to be dedicated both to the real-time and non-real-time traffic on a particular outgoing link in case of congestion. As a consequence, the throughput for normal data does not decrease, provided that this r value is properly adjusted. However, the same r value is initially set for all nodes; the selection is done in such a way that it will satisfy the delay requirement for the least hop node, which does not allow flexible adjustment of bandwidth sharing for different links. Moreover, the average delay increases with a higher real-time data rate. The protocol was extended in Ref. [25] by assigning a different r value for each node to achieve better utilization of the links. In addition, the average delay per packet does not increase overly much with an increase in the real-time data rate. However, finding the r values and sending these to a particular node not only requires overhead but is energy consuming as well, because the r values have to be unicasted to every single node. Moreover, when a route changes, a set of new r values has to be calculated for all the nodes in the new route and transmitted to the nodes. In our protocol, each node locally adjusts the bandwidth and delay requirement based on the path-length and incoming traffic.

1.3 Network Model

WMSNs have several additional features and challenges compared with traditional WSNs. We considered a static wireless network containing multimedia sensor nodes capable of performing all possible application tasks (e.g., capable of sensing video, audio, scalar data). The following definitions describe a network scenario where the proposed routing protocol fits well.

- Multimedia node (M node): A sensor node capable of generating multimedia data (such as video and image) in real-time interaction with the environment is termed an M node. M nodes are often equipped with additional storage capacity to store their local data.
- Scalar node (S node): S nodes are limited to scalar data such as temperature, humidity, and light intensity.

- Sink: A node with higher computational and storage capacity is regarded as the data collection point for the network. For a single sink scheme, the network inherits a many-to-one traffic pattern; all data packets are routed to the sink node.
- Relay node (R node): Due to the limitations of short-range radio connectivity, a network often requires additional nodes to simply relay incoming packets to its next nodes toward the sink.
- Processing hub (PH): Essentially, PH nodes are a subset of R nodes whose main task is some in-network processing (i.e., data aggregation, discard of redundant data) in a distributed fashion in the network. Some PHs are called multimedia processing hubs (MPHs) because of their higher capabilities to process multimedia sensor data.
- Target area: The core purpose of WMSNs is to observe a specified area and report the status in real time. We term this area the *target area*, or *monitored area.*
- Network size: A network is said to have a network size N if there exist a total of N nodes, with the implicit assumption that nodes are uniformly distributed.

Figure 1.1 demonstrates the above components in a single network setup. Both the M nodes and MPH are equipped with a single radio

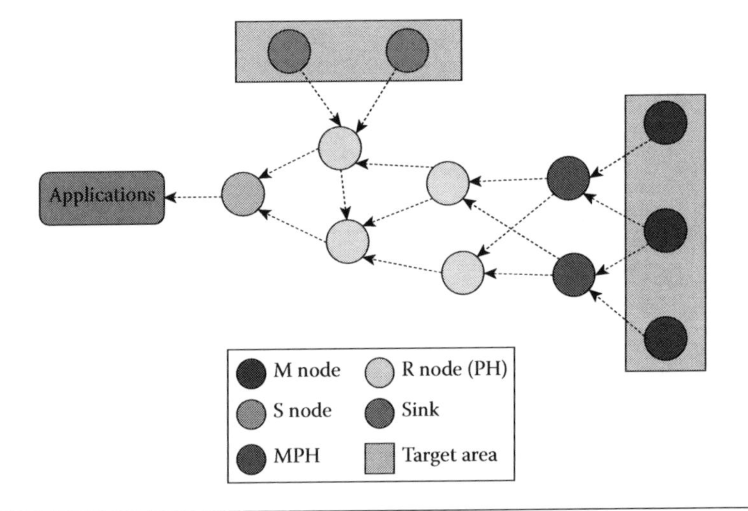

Figure 1.1 Network model: wireless multimedia sensor network.

interface and multiple channels, and the radio interface is capable of transmitting or receiving data on one channel at a given time. The task of the multimedia nodes is to dynamically serve the need of multimedia data to travel from the target area to the sink.

1.4 QoS-Aware Routing Protocol

A QoS-aware routing protocol is presented in this section. In the following, first a multipath, multichannel-provisioned network topology is constructed and, second, a packet-scheduling technique is presented to meet the QoS requirements.

1.4.1 Multipath, Multichannel Network Topology Construction

To realize the network with multipath, multichannel provisions, we used techniques based on a multipath construction mechanism [26] and multichannel assignment technique [9]. The outcome of these techniques was to assign each network node with the knowledge of available paths and channels to transmit and receive data packets. Figure 1.2 shows one possible multipath construction scheme, which has been described [26] as multiple localized, disjoined paths that use localized information alone and do not rely on global topology. As shown in Figure 1.2a, some path request packets have initially been flooded throughout the network by the source nodes. The sink then has some empirical information about which of its neighbors can provide it with the highest quality data (lowest loss or lowest delay).

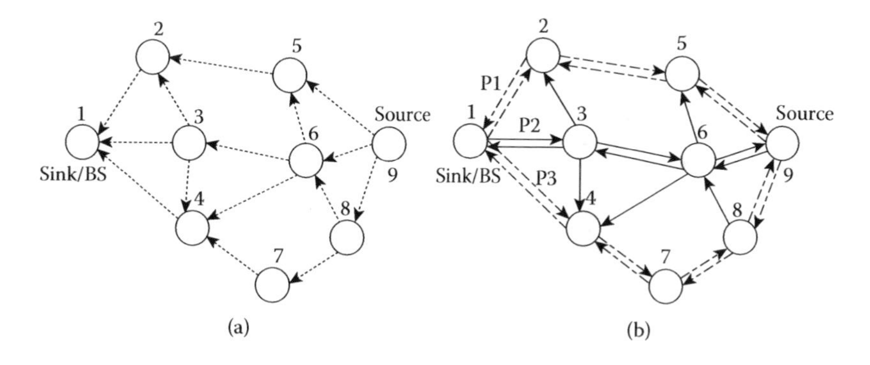

Figure 1.2 Multipath construction: (a) multipath dissemination using flooding (single source to single sink is shown) and (b) three alternative paths P1, P2, and P3 shown with double arrow lines.

To this preferred neighbor, it sends out a primary-path (P3) reinforcement, as shown in Figure 1.2b. As with the basic directed diffusion scheme, that neighbor then locally determines its most preferred neighbor in the direction of the source, and so on. Accordingly, alternative paths P1 and P2 are constructed [26].

The next task is to assign each node its transmission activities to efficiently utilize the bandwidth using multiple frequencies (channels). Mutually orthogonal Latin square (MOLS) based scheduling is applied to assign transmission and reception activities, presented in Ref. [9], as described below.

Definition 1: A $p \times q$ rectangular array formed by the symbols 1, 2, ..., k, where $k \geq p$ and $k \geq q$, is called a Latin rectangle if every symbol from the symbol set appears at most once in each column and once in each row.

Definition 2: A Latin square A of order p is a $p \times p$ matrix with entries from a set of p distinct symbols such that each row and column contains every element exactly once. The symbol in the ith row and the jth column is written as $a_{i;j}$.

Definition 3: Two distinct $p \times p$ Latin squares A and B, where $a_{i,j}$ and $b_{i,j} \in 1, 2, \ldots, p$, are said to be orthogonal if the p^2 ordered pairs $\langle \ldots, (a_{i,j}, b_{i,j}) \ldots \rangle$ are all different.

The square matrices S and R as shown in Figure 1.3 are examples of Latin squares of order 7. In Latin square–based scheduling, channels correspond to the rows and time slots correspond to the columns [27]. According to Definition 3, the two 7 × 7 Latin squares S and R are orthogonal.

Lemma 1: If two nodes are assigned two symbols from two different orthogonal Latin squares, then there is at most one collision for these two nodes in every time frame (proof is given in Ref. [27]).

$$S = \begin{bmatrix} 0 & 1 & 2 & 3 & 4 & 5 & 6 \\ 1 & 2 & 3 & 4 & 5 & 6 & 0 \\ 2 & 3 & 4 & 5 & 6 & 0 & 1 \\ 3 & 4 & 5 & 6 & 0 & 1 & 2 \\ 4 & 5 & 6 & 0 & 1 & 2 & 3 \\ 5 & 6 & 0 & 1 & 2 & 3 & 4 \\ 6 & 0 & 1 & 2 & 3 & 4 & 5 \end{bmatrix} \quad R = \begin{bmatrix} 1 & 2 & 3 & 4 & 5 & 0 & 6 \\ 2 & 3 & 4 & 5 & 6 & 1 & 0 \\ 3 & 4 & 5 & 6 & 0 & 2 & 1 \\ 4 & 5 & 6 & 0 & 1 & 3 & 2 \\ 5 & 6 & 0 & 1 & 2 & 4 & 3 \\ 6 & 0 & 1 & 2 & 3 & 5 & 4 \\ 0 & 1 & 2 & 3 & 4 & 6 & 5 \end{bmatrix}$$

Figure 1.3 Example of two Latin squares of order 7.

During the network initialization phase, a distributed distance-2 vertex coloring algorithm [28] is performed. This approach requires only local information from immediate neighbors to assign the vertex color to the network node. The algorithm outputs different vertex colors to all nodes within interference range of each other (the two-hop distance is a good approximation of the carrier sensing range in ad hoc networks, and node activation scheduling usually requires all neighbors of a node within two hops to be silent when the node transmits [9]). Therefore, the problem of assigning square symbols to network nodes can be modeled as a distance-2 graph coloring problem such that each node can directly use its assigned vertex color as its square symbol. In addition to vertex coloring, MOLS matrices are generated during the network initialization phase.

The orthogonality of the squares corresponds to there being exactly one time/channel assignment for every pair of nodes in different squares. In this way, a node can decide to be a sender or a receiver by picking the appropriate square. For example, let the entries in the square labeled S represent the set of sender nodes and the entries in square R represent the set of receiving nodes. Combining the two squares together will result in a unique time/channel assignment for each pair of senders/receivers, as shown in Figure 1.4. The uniqueness of each pair assignment is guaranteed by the orthogonality of the two squares.

As shown in Figure 1.5, suppose nine nodes numbered from one to nine get seven different colors numbered from zero to six. Any pair of communicating nodes may select appropriate symbols according to their vertex colors to be a sender/receiver pair for a collision-free transmission/reception. With these multipath multichannel provisions,

$$\mathrm{SR} = \begin{bmatrix} (0,1) & (1,2) & (2,3) & (3,4) & (4,5) & (5,0) \\ (1,2) & (2,3) & (3,4) & (4,5) & (5,6) & (6,1) \\ (2,3) & (3,4) & (4,5) & (5,6) & (6,0) & (0,2) \\ (3,4) & (4,5) & (5,6) & (6,0) & (0,1) & (1,3) \\ (4,5) & (5,6) & (6,0) & (0,1) & (1,3) & (2,4) \\ (5,6) & (6,0) & (0,1) & (1,2) & (2,3) & (3,5) \\ (6,0) & (0,1) & (1,2) & (2,3) & (3,4) & (4,6) \end{bmatrix}$$

Figure 1.4 Example of SR matrix.

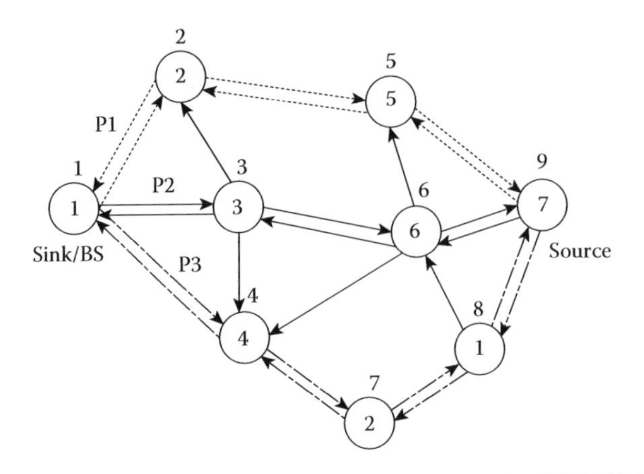

Figure 1.5 Vertex coloring to assign Latin square symbols to nodes. Each circle represents a network node, and assigned colors are numbered inside the circle.

nodes in the network may decide to forward packets to the appropriate path/channel depending on the desired QoS requirements (i.e., required bandwidth and delay as described later).

1.4.2 QoS-Aware Packet Scheduling

In this section, we describe different types of multimedia traffic and queuing models, how to provide QoS assurance, and how to route the traffic toward the destination.

1.4.2.1 Traffic Classification and Queuing Model Sensor data may originate from various types of events that have different levels of importance, as depicted in Table 1.1. Hence, packet-scheduling policy should consider different priorities (importance) for different types of traffic classes. Time-critical (delay bound) packets are assigned a high priority compared to non-time-critical packets to meet the deadlines. Because local packet drop policy is aware of the type of packet it receives, we rationalize it by expanding the length L of each queue q_i a function of its p value. More specifically,

$$L(q_i) = \frac{1}{p\,\text{value}(q_i)} \times \alpha \tag{1.1}$$

where α is a network-dependent constant.

Table 1.1 Multimedia Traffic Classification for Wireless Multimedia Sensor Networks

TRAFFIC CLASS	NAME/EXAMPLE	DELAY	LOSS	BANDWIDTH
Class I ($p = 1$)	Real-time/ video–audio streams	Bounded	Tolerant	High
Class II ($p = 2$)	Real-time/ monitoring processes	Bounded	Intolerant	Low to moderate
Class III ($p = 3$)	Non-real-time/ video–audio stream	Unbounded	Tolerant	High
Class IV ($p = 4$)	Non-real-time/ scalar, snapshot	Unbounded	Tolerant	Low

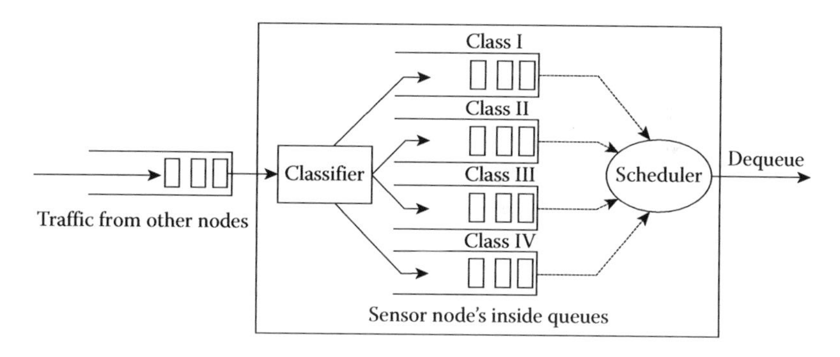

Figure 1.6 Queuing model on a multimedia sensor node.

Figure 1.6 shows the queuing model for a sensor node considering the different traffic classes described in Table 1.1. On each node, there is a classifier to check the type of the incoming packet and to send it to an appropriate queue. Finally, there is a scheduler that schedules the packets according to the delay bound and bandwidth requirements. Note that Equation 1.1 ensures that a high priority packet is assigned to a longer queue and vice versa.

1.4.2.2 QoS Assurance To meet the QoS requirements for a packet from source to destination along a path, let us derive a path-specific condition. Suppose a packet p_i on path P originated at the source node at time t_i and has to reach the destination by $t_i + T_i$, where T_i is the deadline of packet p_i. The arrival time of the packet at hop

j denotes the time it is inserted into the queue at that node. The departure time of p_i from hop j denotes the time the transmission of p_i is completed. The arrival time of p_i at hop $j+1$ is equal to its departure time from j plus propagation delay. Let the time this packet p_i spends at hop j be d_j, which is the interval between its arrival time and departure time at hop j; let s_j denote the switching delay from one channel to another at each hop. So, the packet p_i will reach the destination while preserving the delay bound if

$$\sum_{j=1}^{H} d_j + \rho_j + s_j \le T_i \tag{1.2}$$

where ρ_j is the propagation delay for each hop j and H is the total number of hops a packet travels. The propagation delay in Equation 1.2 can be neglected, since packet propagation occurs at the speed of light and is therefore much faster than transmission and queuing delays. Considering the delay for a specific path, the sink (network designer) may determine the required bandwidth consumed by different traffic classes. We denote B as the required bandwidth. Initially, the sink will determine the value of B based on the observed delay for a time-critical traffic class and will broadcast this value. After receiving the value, all nodes will dynamically calculate their own value for B considering the distance of the nodes from the sink for a particular path. Then a PPDD model will determine the delay encountered by each packet in a particular queue along the path. Finally, the waiting time priority (WTP) algorithm [12] will be exploited to dequeue packets from the queues according to the service class and waiting time.

Dynamic Bandwidth Adjustment As both real-time and non-real-time traffic coexist, bandwidth should be used effectively, so that not only are the QoS requirements of real-time traffic met but service to the non-real-time traffic is also maximized. As mentioned earlier, a parameter B is used to control the bandwidth used by real-time and non-real-time traffic. As shown in Figure 1.7, Node 4 has more traffic than Node 3, 2, or 1; accordingly, Node 4 should allocate more bandwidth. We assume that the rate of real-time data is almost inversely proportional to the hop count of the node from the sink.

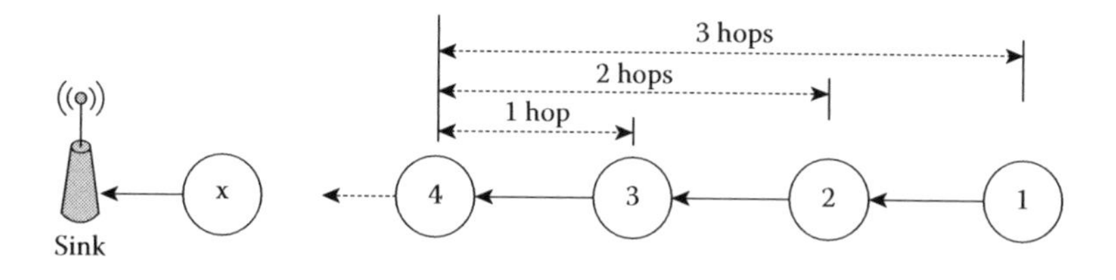

Figure 1.7 Dynamic bandwidth adjustment and path-length-based proportional delay differentiation calculation.

As stated earlier, if the sink broadcasts a value for B based on the observed delay for time-critical traffic, then each node adjusts its own B value according to the following equation:

$$B_H = \frac{B}{H + \alpha} \tag{1.3}$$

Here, H is the hop count of the node from the sink; α is the adjusting factor and the node will set this value based on the incoming traffic. When the sink observes that the end-to-end delay is increasing, it increases the value of B to allocate more bandwidth to real-time traffic, and vice versa. In addition, the sink tries to make the value of B as low as possible without violating the QoS requirement to maximize the bandwidth use of non-real-time traffic.

PPDD: The PPDD is a differentiated services based proposal that is an extension of PDD [12], defined for wired networks. The PPDD scheduler services packets in classes and realizes proportional average per-hop queuing delays among them locally at each node along the path. At node k, packets from class i experience smaller delay than class j for all $i > j$, $i, j \in S_{b,k}$, where $S_{b,k}$ is the set of backlogged classes at node k. Usually, the end-to-end delay of a packet is proportional to the number of hops. For example, for a path, a packet that is H hops away from the sink experiences a smaller end-to-end delay than a packet that is more than H hops away from the sink. As shown in Figure 1.5, a packet from Node 1 will experience more end-to-end delay than a packet from Node 4. The spacing between delays is tuned by the sink based on observed delay of the real-time packet with a set of class differentiation parameters. As its name suggests, the model not only holds at each node it also holds across all nodes in a path. The PPDD service model is defined as follows.

Let $1 = \partial_1 > \partial_2 > \ldots > \partial_H > 0$ be delay differentiation parameters that define that a packet of a node smaller hops away from the sink may allow higher delay than a packet that arrives from a node more hops away. Let d_H^k denote the average queuing delay of a packet at node k that is H hops away from the sink. Then, the PPDD requirement is given according to the following equation:

$$\frac{d_H^k}{d_{H+1}^k} = \frac{\partial_H}{\partial_{H+1}} \tag{1.4}$$

Then, with WTP, each class is serviced with a separate first-in-first-out queue. The head-of-line packet of a class is assigned a WTP based on the service class and waiting time of the packet. The scheduler always schedules the highest priority head-of-line packet for transmission.

1.4.2.3 Routing *Single Sink* The packets are routed through the nodes along the path from the source to the destination; nodes choose the paths/channels that meet the bandwidth and delay requirements. Each node knows the available path options and collision-free channel assignment among its two-hop neighbors and adjusts bandwidth and delay according to Equations 1.2 and 1.3, respectively, to relay traffic along the path. Packets that do not meet the deadline (i.e., QoS requirements) are discarded. Best-effort traffic is routed through the alternative paths to balance the distribution of the remaining traffic. Redundant data are aggregated by the PHs to reduce the network traffic.

Multiple Sinks Initially, we considered deployment of WMSNs based on a many-to-one communication paradigm, where a single sink collects data from a number of data sources. Because our protocol deals with real-time and non-real-time data, we may exploit scenarios with multiple sinks to further balance the distribution of traffic. With the resulting many-to-many communication paradigm, each node might adjust the bandwidth, delay, and path length of different sinks to route the packets. For example, real-time data may be routed to the nearest sink (i.e., with a path length smaller than that of other available sinks), and less sensitive non-real-time data may be routed to the longest route, because the delay requirement is flexible for such data.

1.5 Performance Evaluation

The effectiveness of the proposed QoS routing approach was evaluated through simulation in ns-2 [29] under various situations. Initially, we considered a network of size 100 (N = 100) uniformly placed in a 1000 × 1000 meter area. Nodes were positioned as a 10 × 10 square grid. Nodes within one and two hops were marked as R nodes. The sink was placed at the center (0,0). Other nodes were operated as either M nodes or N nodes. Table 1.2 shows the simulation parameters in more detail. Some parameters of the table were taken from Ref. [25]. To evaluate the performance of the proposed routing approach, we used the following performance metrics, which apply to the entire evaluation section:

- EDP (end-to-end delay per packet): The EDP is measured as the time difference between sensing the data and receiving it by the sink.
- PDR: The PDR indicates the number of packets dropped per time unit due to congestion or local buffer overflow.
- NDR: The NDR is computed as the ratio of the total number of successfully delivered packets to the total number of packets sent by all source nodes in the network [30].
- Network throughput: The network throughput is measured as the total number of data packets received at the sink divided by the entire simulation time.

Table 1.2 Simulation Parameters

PARAMETERS	VALUE
Network size (N)	100
Radio model*	Free space
Packet generation rate (non-real-time)*	1 packet/s
Packet generation rate (real-time)*	8 packet/s
Maximum data packet length*	10 kbit
Maximum control packet length*	2 kbit
Total number of channels	7
Channel switching delay	250 ms
Number of M nodes	15 (15% of N)
Number of S nodes	15 (15% of N)
Simulation duration	100 s
Number of iterations	10

1.5.1 *Effects on End-to-End Delay*

First, we consider the impact of the real-time data rate on the average delay per packet for both real- and non-real-time data. The average delay per packet is defined as the average time a packet takes to travel from a sensor node to the sink. We observed that both multiple-*r* and single-*r* mechanisms had a higher average delay compared to our proposed protocol, as shown in Figure 1.8a. The multiple-*r* mechanism performed better than the single-*r* mechanism, as expected; every particular node adjusts its *r* value based on the resources available. This method is more efficient than the single-*r* mechanism, in which a unique *r* value is imposed by the sink for all the nodes. Intuitively, the average delay per packet for the proposed protocol is less than for the single-*r* mechanism. The reason is that forwarding nodes locally adjust the bandwidth (value of *B*) proportional to the expected load instead of assigning a single value of *r* for all forwarding nodes. Moreover, even though the value of *B* is not set exactly like the multiple-*r* mechanism, the average delay is less for the multiple-*r* mechanism. The rationale behind this system is that the multiple-*r* mechanism uses unicast transmission to deliver the individual value to the nodes; our protocol requires smaller control packets (because a single *B* value is sent to all nodes), which in turn increases the forwarding rate of the data packets.

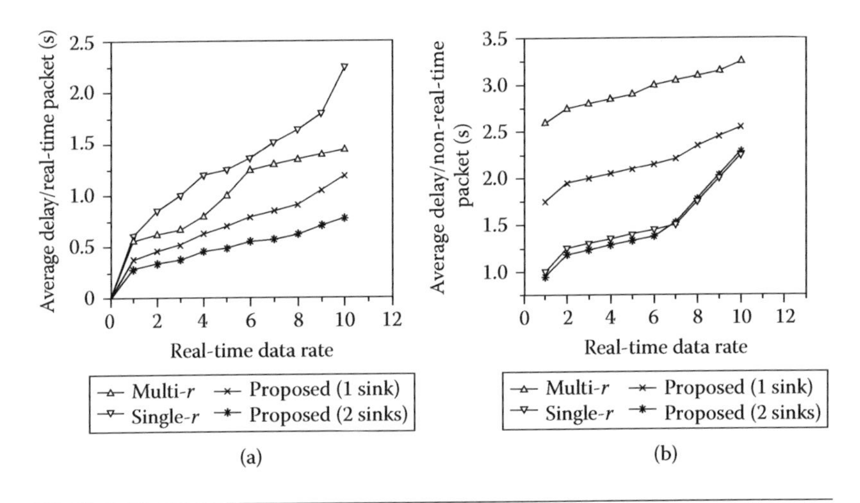

Figure 1.8 Impact of the real-time data generation rate on the (a) average delay per real-time packet and (b) average delay per non-real-time packet.

Figure 1.8b shows the effect of the real-time data rate on the average delay per non-real-time packet. The delay increases with the rate, because packets incur more queuing delay and share the same amount of bandwidth. Note that the average delay for non-real-time packets using the multiple-*r* mechanism is greater than for the single-*r* mechanism. With the multiple-*r* mechanism, the increase in the throughput of non-real-time packets causes an extra queuing delay on the nodes, leading non-real-time packets to experience end-to-end delay [25]. Our protocol has less average delay compared to the multiple-*r* protocol, because the nodes can schedule non-real-time packets and exploit multiple paths.

1.5.2 Scalability Assessment

To evaluate the scalability of the proposed routing scheme, we repeated the simulation with different settings for the network size (N). In this case, we varied N from 25 to 200. Note that the end-to-end delay with a smaller network ($N \leq 50$) was not significantly different for either the previous or proposed schemes (Figure 1.9). However, with a larger network, the improvement was remarkable; for instance, a network with $N = 200$ using our proposed routing scheme (with one source) experienced an average of 47.8% less delay compared with that of the previous method using the single-*r* mechanism. Apart from this advantage, the presented methods showed a linear increase in their end-to-end delay parameters with respect to network size. Therefore, the routing mechanism presented herein can be efficiently deployed for a larger network.

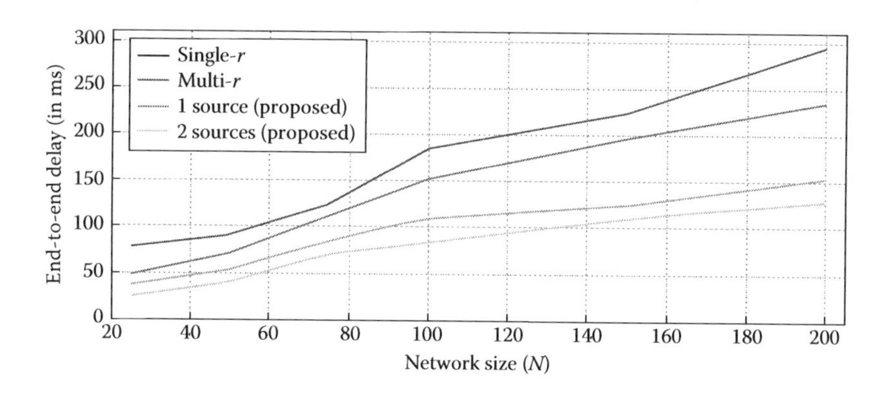

Figure 1.9 End-to-end delay for a large network.

1.5.3 Effects on Network Throughput

Next, we consider network throughput, which is measured as the total number of data packets received at the sink divided by the simulation time. Figure 1.10a shows the non-real-time data throughput. Our protocol outperforms existing protocols due to efficient utilization of the wireless spectrum. When the number of real-time packets increases, it gets more difficult to satisfy the increasing need for QoS paths, leading to rejection of paths or packet drops for non-real-time data and causing throughput for such data to decrease. Figure 1.10b shows the average lifetime of the nodes, illustrating that our model consumes less energy compared to both the other models. The reason is that our model does not require multiple unicast transmission of the r value, unlike the multiple-r mechanism. Moreover, the PHs perform in-network data aggregation and channel assignment, which results in fewer collisions.

1.5.4 Packet Drop Ratio

One of the major motivations for this work stems from the fact that, in multi-application WSN setups, it is not wise to treat every data packet equally, as most routing protocols do [30]. In this section, we analyzed the trace from the simulation to assess the effects of PDR on each type of traffic (Classes I to IV). Figure 1.11 depicts the findings. It shows that

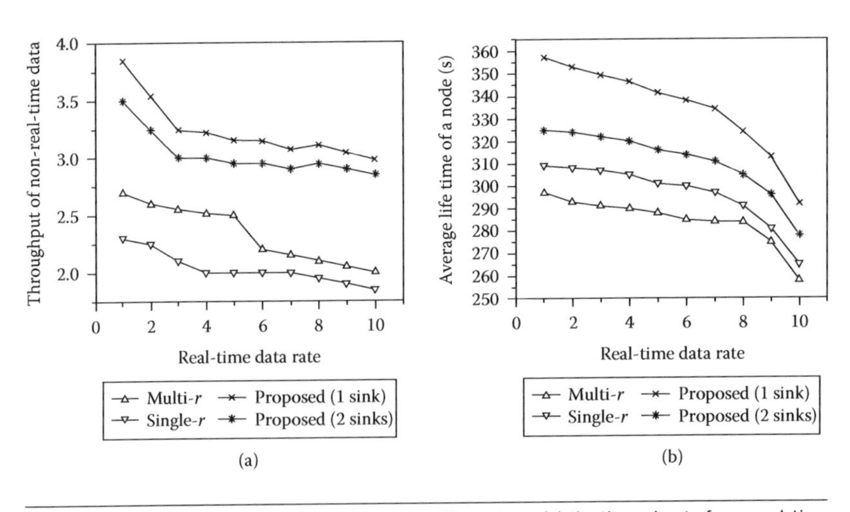

Figure 1.10 Impact of the real-time data generation rate on (a) the throughput of non-real-time data and (b) the average lifetime of a node.

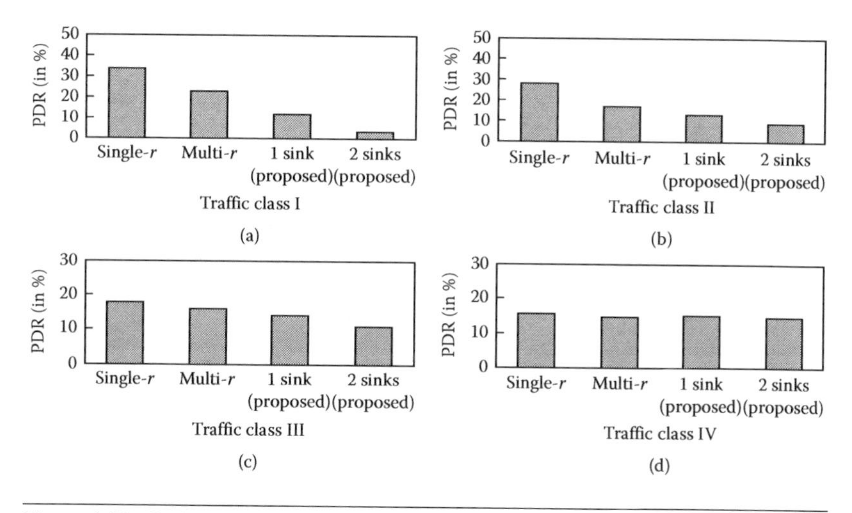

Figure 1.11 Packet drop ratio: a comparative study with different classes of data traffic.

for traffic of Classes I, II, and III, the traditional methods (multiple-*r* and single-*r* mechanisms) incurred a higher PDR compared with our proposal. This higher rate is a result of the prioritization of each packet. High priority packets are assigned to longer queues, which ensures lower probability of being dropped because of local buffer overflow. Furthermore, they are transmitted in a multipath fashion to attain higher network reliability. In the case of low priority packets (Class IV), the achievement is not noticeable, because these packets are the primary targets to be dropped. We occasionally obtained a higher PDR using our proposal (data not shown; only the averaged value is plotted in Figure 1.11).

1.5.5 Network-Wise Delivery Ratio

The primary goal of this work is to ensure higher reliability in delivering sensor information for MWSNs. We assessed the NDR to justify the network-wise overall improvement. We varied the network size N from 25 to 200. The findings are graphed in Figure 1.12. For smaller networks, the NDR was almost 100% for all schemes. With multiple-*r* and single-*r* schemes, the NDR dropped rapidly when N became larger than 50: the average NDRs recorded were 81.67% and 83.78% for the single-*r* and multiple-*r* schemes, respectively. Using our proposed schemes, the average NDRs were 92.7% and 95.167% for one- and two-sink networks, respectively. These figures suggest that our proposed scheme ensures a high degree of reliability in terms of NDR even for a very large network.

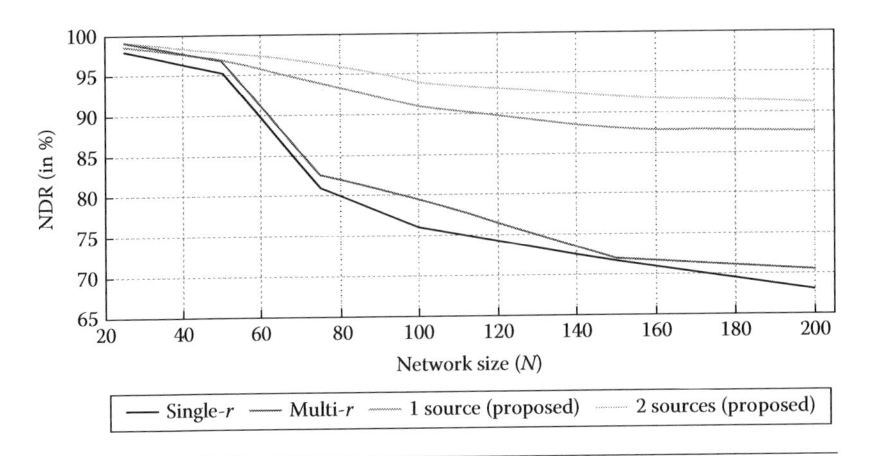

Figure 1.12 Network-wise delivery ratio: a comparative study.

1.6 Conclusions

In this chapter, we described and presented a QoS-aware routing mechanism to meet the challenges posed by WMSNs. We considered many-to-one and many-to-many communication scenarios to evaluate the performance of the scheme using a single sink and multiple sinks (two sinks). In all cases, the sensor nodes were considered to be static. We showed that the proposed QoS-aware routing mechanism provides a significant performance improvement in terms of average delay, average lifetime, network throughput, PDR, and NDR. In particular, we showed that, with more than one sink, data gathering can be further improved with the proper routing technique. In future works, it would be interesting to experiment with performance issues for the mobile sink and at the same time to implement this protocol in a test-bed to measure its efficiency with commercial sensors.

References

1. Szewczyk, R., Mainwaring, A., Polastre, J., Anderson, J., Culler, D. An analysis of a large scale habitat monitoring application. In: *Proceedings of the 2nd International Conference on Embedded Networked Sensor Systems, SenSys '04*, pp. 214–226. ACM, New York, NY, 2004. DOI: 10.1145/1031495.1031521.
2. Tolle, G., Polastre, J., Szewczyk, R., Culler, D., Turner, N., Tu, K., Burgess, S. et al. A macroscope in the redwoods. In: *Proceedings of the 3rd International Conference on Embedded Networked Sensor Systems, SenSys '05*, pp. 51–63, ACM, New York, NY, 2005. DOI: 10.1145/1098918.1098925.

3. Ingelrest, F., Barrenetxea, G., Schaefer, G., Vetterli, M., Couach, O., Parlange, M. Sensorscope: Application-specific sensor network for environmental monitoring. *ACM Transactions on Sensor Networks* 6, 1–32, 2010. DOI: 10.1145/1689239.1689247.
4. Tavakoli, A., Kansal, A., Nath, S. On-line sensing task optimization for shared sensors. In: *Proceedings of the 9th ACM/IEEE International Conference on Information Processing in Sensor Networks, IPSN '10*, pp. 47–57. ACM, New York, NY, 2010. DOI: 10.1145/1791212.1791219.
5. Bhattacharya, S., Saifullah, A., Lu, C., Roman, G.C. Multi-application deployment in shared sensor networks based on quality of monitoring. In: *Proceedings of the 2010 16th IEEE Real-Time and Embedded Technology and Applications Symposium, RTAS '10*, pp. 259–268, IEEE Computer Society, Washington, DC, 2010. DOI: 10.1109/RTAS.2010.20.
6. Akyildiz, I.F., Melodia, T., Chowdhury, K.R. A survey on wireless multimedia sensor networks. *Computer Networks* 51(4), 921–960, 2007. DOI: 10.1016/j.comnet.2006.10.002.
7. Data Sheet for Micaz Mote. http://www.xbow.com/Products/wCatalogs.aspx, 2010.
8. Kyasanur, P., Vaidya, N.H. Capacity of multi-channel wireless networks: Impact of number of channels and interfaces. In: *Proceedings of ACM International Conference on Mobile Computing and Networking (MobiCom)*, pp. 43–57, Cologne, Germany, 2005.
9. Alnifie, G., Simon, R. A multi-channel defense against jamming attacks in wireless sensor networks. In: *Proceedings of the 3rd ACM Workshop on QoS and Security for Wireless and Mobile Networks Chania*, pp. 95–104, Crete Island, Greece, 2007.
10. Le, H.K., Henriksson, D., Abdelzaher, T. A control theory approach to throughput optimization in multi-channel collection sensor networks. In: *Information Processing in Sensor Networks, 2007. IPSN 2007. 6th International Symposium*, pp. 31–40, ACM: New York, NY, 2007. DOI: 10.1109/IPSN.2007.4379662.
11. Hamid, M.A., Alam, M.M., Hong, C.S. Design of a QoS-aware routing mechanism for wireless multimedia sensor networks. In: *IEEE Global Communications Conference (GLOBECOM)*, pp. 800–805, 2008.
12. Dovrolis, C., Stiliadis, D., Ramanathan, P. Proportional differentiated services: Delay differentiation and packet scheduling. *IEEE/ACM Transactions on Networking* 10(1), 12–26, 2002.
13. Gurses, E., Akan, O.B. Multimedia communication in wireless sensor networks. *Annales des Telecommunications* 60(7–8), 872–900, 2005.
14. Ehsan, S., Hamdaoui, B. A survey on energy-efficient routing techniques with QoS assurances for wireless multimedia sensor networks. *IEEE Communications Surveys Tutorials* 14(2), 265–278, 2012.
15. Cucchiara, R. Multimedia surveillance systems. In: *VSSN '05: Proceedings of the Third ACM International Workshop on Video Surveillance & Sensor Networks*, pp. 3–10, ACM, New York, NY, 2005. DOI: 10.1145/1099396.1099399.

16. Mao, S., Bushmitch, D., Narayanan, S., Panwar, S. MRTP: A multiflow real-time transport protocol for ad hoc networks. *IEEE Transactions on Multimedia* 8(2), 356–369, 2006.
17. Yang, X., Vaidya, N.H. A wakeup scheme for sensor networks: Achieving balance between energy saving and end-to-end delay. In: *Proceedings of IEEE RTAS*, 10th IEEE, pp. 19–26, 2004. DOI: 10.1109/RTTAS.2004.1317245.
18. Shah, R., Rabaey, J. Energy aware routing for low energy ad hoc sensor networks. 2002. citeseer.ist.psu.edu/shah02energy.html
19. Hamid, Z., Bashir, F. XL-WMSN: Cross-layer quality of service protocol for wireless multimedia sensor networks. *EURASIP Journal on Wireless Communications and Networking* 2013(1), 1–16, 2013.
20. Kim, K.I., Sung, T.E. Modeling and routing scheme for (m, k)-firm streams in wireless multimedia sensor networks. *Wireless Communications and Mobile Computing* 15(3), 475–483, 2013.
21. Shah, G., Liang, W., Akan, O. Cross-layer framework for QoS support in wireless multimedia sensor networks. *IEEE Transactions on Multimedia* 14(5), 1442–1455, 2012.
22. Touil, H., Fakhri, Y., Benattou, M. Energy-efficient MAC protocol based on IEEE 802.11e for wireless multimedia sensor networks. In: *2012 International Conference on Multimedia Computing and Systems (ICMCS)*, pp. 53–58, 2012.
23. Felemban, E., Lee, C.G., Ekici, E. MMSPEED: Multipath multi-speed protocol for QoS guarantee of reliability and timeliness in wireless sensor networks. *IEEE Transactions on Mobile Computing* 5(6), 738–754, 2006.
24. Akkaya, K., Younis, M. An energy-aware QoS routing protocol for wireless sensor networks. In: *Proceedings of the MWN*, pp. 710–715, 2003. DOI: 10.1109/ICDCSW.2003.1203636.
25. Akkaya, K., Younis, M. Energy and QoS aware routing in wireless sensor networks. *Cluster Computing* 8(2–3), 179–188, 2005. DOI: 10.1007/s10586-005-6183-7.
26. Ganesan, D., Govindan, R., Shenker, S., Estrin, D. Highly-resilient, energy-efficient multipath routing in wireless sensor networks. SIGMOBILE Mob, *Comput. Commun. Rev.* 5(4), 11–25, 2002.
27. Ju, J., Li, V.O.K. TDMA scheduling design of multihop packet radio networks based on Latin square. *IEEE Journal on Selected Areas in Communications* 17(8), 3499–3504, 1999.
28. Kubale, M., Kuszner, L. A better practical algorithm for distributed graph coloring. In: *PARELEC '02: Proceedings of the International Conference on Parallel Computing in Electrical Engineering*, p. 72. IEEE Computer Society, Washington, DC, 2002.
29. The Network Simulator—ns-2. http://www.isi.edu/nsnam/ns/index.html, 2010.
30. Puccinelli, D., Haenggi, M. Reliable data delivery in large-scale low power sensor networks. *ACM Transactions on Sensor Networks* 6, 28:1–28:41, 2010. DOI: 10.1145/1777406.1777407.

2

Adaptation Techniques for Multimedia Communication in Wireless Sensor Networks

PINAR SARISARAY-BOLUK

Contents

2.1 Introduction

New advances in (complementary metal-oxide semiconductor) CMOS technology have led to an increase in the availability of cameras and microphones. Wireless multimedia sensor networks (WMSNs) have gained attention after integration of this multimedia-enabled low-cost hardware into traditional sensors [6,12,31]. WMSNs are the next-generation WSNs; they are specially designed to retrieve and transmit multimedia data from wireless environments. They support various multimedia applications including multimedia surveillance networks, target classification, disaster prevention and environmental monitoring, and many more. A comprehensive survey of multimedia communication over WMSNs is presented in Refs. [5,36]. Real-time data transmission over the wireless medium has rigid quality-of-service (QoS) needs; several requirements based on multimedia applications are also given in Refs. [7,25]. Traditional sensor nodes are not suitable for multimedia applications due to their exhaustive requirements. Instead, various improved series of wireless sensor nodes such as Imote2 [26], TelosB motes [28], and MICAz [Crossbow technology, California, www.xbow.com] may be utilized for multimedia applications.

Figure 2.1 shows several new imaging platforms for sensor motes, specifically the CMUcam3 [Carneige Mellon University, Pittsburgh] [31], Cyclops [Agilent Laboratories and the Center for Embedded Network Sensing at UCLA, Los Angeles] [29], and Stargate board [Crossbow technology, California] with webcam [15]. The Crossbow Stargate platform provides a processing platform that can plug into a webcam for medium-resolution imaging. It is utilized as a video sensor node in WMSNs. The Cyclops may be connected to a sensor node such as Crossbow's Mica2 or MICAz [44] that provides low-resolution imaging. It also contains several special image processing libraries. The CMUcam3 is a recent series of embedded cameras that can plug into the TMote Sky. It is composed of the OmniVision CMOS camera sensor module and open-source libraries.

(a) (b)

(c)

Figure 2.1 Several recent imaging platforms for sensor motes.

These libraries include several image processing algorithms such as frame differentiae, JPEG compression, histogramming, and edge detection, as presented in Figure 2.1.

Construction of a WSN for a specific application is affected by various considerations such as scalability, operating environment, network topology, production costs, hardware constraints, transmission media, and energy consumption [4]. There are several additional factors that affect the performance of WMSNs, for example, multimedia coding techniques, high bandwidth demand, and application-specific QoS requirements. Because sensor nodes have limited power resources, storage, computation, and communication capability, it is challenging to fulfill the application layer QoS requirements for multimedia communication over lossy wireless links. To transmit multimedia data effectively, these factors should be considered in the design of WMSN protocols [16,33].

They are explained in more detail as follows:

- Application-specific QoS requirements: Due to the nature of the wireless medium, multimedia traffic is exposed to losses during transmission, which leads to perceptual quality degradation for viewers at the sink. Furthermore, applications generally require that transmission of multimedia data be completed in a certain time period, referred to as a *delay bound*. This time period is composed of processing time and communication latency. Hence, protocols should be designed to fulfill the reliability and delay requirements of multimedia transmission.

- High bandwidth demand: Most multimedia applications require a certain bandwidth to provide the perceptual quality and delay bound required by the application layer. Otherwise, the application layer QoS requirements cannot be satisfied by the network. Hence, solutions for meeting high bandwidth requirements should be considered in the design of WMSN protocols [7,25,37].
- Error control: To decrease the effects of data loss on multimedia data, error-mitigating techniques are utilized in coded video [1]. However, the perceptual quality of multimedia data is still very sensitive to losses due to the lossy environment and sensor node failures. Hence, error mitigation techniques should be analyzed and designed for WMSNs so as to decrease quality degradation during communication [8].
- Packet prioritization: Multimedia data may be composed of valuable parts and parts that contain less information. Losses of the valuable parts of multimedia data lead to more perceptual quality degradation at the end-user. In order to protect the more valuable parts during transmission, multimedia packets should be prioritized with respect to their significance [10].

There are only a few studies related to satisfying multimedia requirements in WMSN. Many of the proposed protocols do not consider the real-time requirements of applications or only try to meet the deadlines as fast as possible [3,14,17]. Few studies take into account the nature of the multimedia data during communication in WMSNs [19,34]. They generally ignore the transmitted data content, which plays a big role in meeting requirements of the application. Due to this lack of awareness, WMSNs cannot employ the most appropriate transmission techniques to present the needed quality for the application layer. It is vital for QoS-based WMSN applications to provide accurate interaction between the content of the multimedia data and communication protocols. In this chapter, detailed information is provided about well-known existing communication protocols from the media access control (MAC) layer to the transport layer for WSNs. However, these communication protocols were not designed for real-time multimedia communication; therefore, each underlying communication protocol must be adapted to fulfill the requirements of the applications. To satisfy QoS-based WMSN applications, the characteristics of the multimedia data are integrated with the communication architecture of the sensor networks. To remedy this situation, we describe the weaknesses of the existing protocols. We then provide some adaptation techniques to make them suitable for multimedia communication.

The rest of this chapter explains WSN protocols and algorithms and evaluates them in terms of multimedia transmission. Some multimedia adaptation techniques are described for each protocol to provide a better perspective for future deployment. Section 2.2 presents transport layer protocols. Section 2.3 provides network layer protocols that are essential to sensor network applications. Section 2.4 illustrates MAC layer protocols. Finally, Section 2.5 contains our conclusions.

2.2 Transport Layer Protocols

Transport protocols are a means of dealing with reliability and congestion, supporting fair bandwidth sharing, and assuring end-to-end (E2E) reliability. Although these objectives are still absolute for multimedia applications, additional vital issues must be addressed in these protocols. Transport layer protocols should meet real-time requirements such as delay, jitter, bandwidth, etc. In this section, we consider several transport protocols for WSN and check their suitability for WMSN applications. Additionally, several solutions are given to increase the suitability of these protocols for multimedia applications.

2.2.1 Event-to-Sink Reliable Transport Protocol

The event-to-sink reliable transport (ESRT) algorithm [2,39] maintains reliable event detection without intermediate caching requirements. The most important characteristics of ESRT are its self-configuring ability, energy attentiveness, and congestion control. ESRT utilizes both the observed reliability, which is the number of packets that are passed from the event to the sink, and the required reliability, which is the desired number of such packets for the event to be successfully tracked to adjust reporting frequency. If the observed reliability of the event is less than the requirement, then ESRT increases the reporting frequency. On the contrary, if the required reliability degree has been transcended, ESRT reduces the reporting frequency to preserve energy. The frequency at which sensors must send their reports is conveyed to them through broadcasts from the sink, after appropriate calculations, so that the necessary reliability is achieved. Congestion control is performed by controlling buffer levels at forwarding sensors. In this protocol, a sensor node does not require a sensor ID because it uses the event ID for the communication. The advantage of ESRT algorithms is that the protocol is mainly executed in the sink with a plentiful energy. Minimal functions of the protocol run on resource-constrained sensor nodes. This protocol has some limitations in supporting multimedia data.

Limitations:

- E2E QoS [11]: Multimedia applications have different QoS requirements so as to provide user-level satisfaction in WMSNs. Hence, the QoS parameters traditionally used may not be adequate for these applications. Several new QoS metrics are needed for effective evaluation of the transmission of real-time data.

  The QoS requirements of WMSN multimedia applications may vary, and traditional E2E QoS parameters may not be sufficient to portray WMSN application-specific QoS requirements. Consequently, new QoS parameters are required for measuring the delivery of multimedia data in an efficient and effective way [9].

- Inability to cope with bursty multimedia traffic: ESRT mainly considers reliability and energy conservation, without paying enough attention to congestion. Transmission of especially large amounts of multimedia data may lead to congestion over the network. Congestion consumes the limited energy as a result of a large number of retransmissions and packet losses. It precludes event discovery reliability [41].

- Weak algorithm for high bandwidth applications [25]: Multimedia data represent a high volume of data. Hence, it is necessary to fragment the data into smaller packets.
- Rate control: ESRT presents reliability on an application level. However, it does not afford real-time data-dependent source rate managing.

Solutions:

- Because ESRT protocol is based on events, many sensors may be involved in transmitting high-volume multimedia traffic. An event report may be divided into smaller packets coming from the source. Thus, it requires a redefinition of the observed reliability to examine the number of obtained reports taken by the sink instead of the packets. Before transmission of this traffic, a filter may be required to reduce the amount of data to be sent. Additionally, a smarter algorithm regulating the data rate of the source is needed to handle congestion and jitter because of the nature and size of multimedia traffic. The algorithm should control the reporting frequency adaptively.

2.2.2 *Pump Slowly, Fetch Quickly: A Reliable Transport Protocol for Sensor Networks*

The reliable event-to-sink communication necessary for the delivery of data from the sink to sensors contains retransmission of the packets as well as an acknowledgment procedure, which drains sensor network resources. Hop-by-hop retransmission and Negative Acknowledgment (NACK) mechanisms are generally superior to acknowledgment (ACK) mechanisms and E2E retransmissions in providing energy efficiency. The sink, with abundant energy, storage, and processing and communication resources, takes a greater share of responsibility in the sink-to-sensor communication on the reverse path by using a powerful antenna. Hence, data communication along this path may be subject to less congestion than the forward path, which is based on multihop communication.

Less congestion leads to decreased usage of the congestion-control mechanisms on the reverse path in WSN. Pump slowly, fetch quickly (PSFQ) [38] is a reliable, robust, scalable, and customizable transport protocol. It consists of three operations: message relaying (pump), error recovery (fetch), and selective status reporting (report). The pump function delivers data to the sensor nodes while controlling flow; it localizes loss by guaranteeing buffering of data packets at intermediate nodes. Thus, the errors on one link are rectified locally without propagating them down the entire path. Intermediate sensor nodes send the packet by considering loose delay bounds. Whenever a receiver discerns gaps in the received sequence numbers, a loss is denoted, and it goes into fetching mode. PSFQ deduces that data loss results from unsatisfactory link conditions rather than traffic congestion. It uses a link-based error correction scheme, which causes overhead because it requires intermediate nodes to become aware of a forwarded packet. It requests missing packets to be sent from neighboring nodes. An attempt is made to aggregate losses—that is, several message losses are batched into a single fetch operation, which is especially appropriate for burst losses. PSFQ performs a reporting operation to assure feedback on packet transmission status to the source. The farthest target sensor initiates its report on the reverse path of data, and all intermediate nodes append their reports to the same report. Hence, PSFQ guarantees that data segments are delivered to all necessary receivers in a scalable and reliable manner, even in environments with poor channel conditions.

Limitations:

- Congestion is unaccounted for: This protocol attempts to restrict congestion by injecting packets more slowly. However, the delivery of multimedia data to the network may lead to congestion. The huge volume of multimedia data quickly drains sensor resources as well as network bandwidth. It results in severe packet losses and node failures. Although PSFQ lacks the ability to differentiate the reason for packet losses, it presumes that the packet losses occur due to channel problems without considering congestion. As a result, there is no operation to handle congestion problems in this protocol.
- Slow pump operation and buffering cause large delays: As slow pumping causes a delay at intermediate hops,

the WMSN suffers from high latency. However, the transmission of the multimedia data should be completed in a certain time period. This time period includes the processing time and communication latency.

- Hop-by-hop error correction (recovery) with cache requires more buffers: When the receiver realizes that there is a missing packet, it caches received packets until the lost one is taken. This operation introduces overhead at the receiver node, which has limited buffer storage.

Solutions:

- This algorithm does not function adequately, because all losses are reported as channel losses. This misclassification decreases the performance of PSFQ over wireless channels for multimedia transmission. A new algorithm that includes loss differentiation and source rate adjustment should be implemented to differentiate channel and congestion status.

2.2.3 CODA: Congestion Detection and Avoidance in Sensor Networks

CODA's [40] goal is to provide no protocol overhead throughout normal network conditions. However, it should react quickly to alleviate congestion around hotspots, whenever congestion is noted. Congestion detection, open-loop hop-by-hop back pressure, and closed-loop E2E multisource regulation are three functions included in the protocol. In CODA, congestion is determined by monitoring channel and buffer usage. If buffer and channel load are greater than a predefined value, it indicates congestion. In case of congestion, upstream nodes are alerted to lessen their transmission rate by way of open-loop hop-by-hop back pressure. By means of this mechanism, upstream nodes reduce their transmission rate. These nodes may then propagate the back-pressure upstream depending on the local congestion status. CODA can also adjust the multisource rate using a closed-loop E2E approach. With this approach, when the sensor rate is greater than the theoretical throughput, the sensor sets a regulation bit in an event packet. If the sink receives an event packet with a regulation bit, it transmits an ACK message to the other sensors to reduce their rate. When the congestion is cleared, another ACK message is delivered to the nodes to notify them to increase their rate. CODA has some limitations such as

Limitations:

- Design avoids reliability: This algorithm considers not reliability but congestion control. It tries to prevent congestion, which leads to an increase in the performance of network. However, it does not consider reliability.
- Very low link utilization: Wireless channels may not allow a high data rate during monitoring of the event.
- Resource wastage problem: The suppression messages and ACK mechanism cause waste of resources such as energy and bandwidth.
- Experienced delay: The reaction time for closed-loop multisource regulation increases with excessive congestion because there is a high probability of loss for the ACK coming from the sink.
- Weak algorithm for Real Time (RT) and high bandwidth applications: CODA cannot cope with bursty multimedia traffic at intermediate nodes due to queue lengths.

Solutions:

- A new multipath algorithm should be implemented for reliability purposes. It should also cope with congestion problems by splitting a large data burst into smaller ones.

2.2.4 Sensor Transmission Control Protocol

Sensor Transmission Control Protocol (STCP) is a reliable and robust transport layer protocol whose functions are generally performed by the sink. It provides regulated reliability and congestion detection-avoidance mechanisms for multiple network applications. In this protocol, first the sensor nodes send a session initiation packet to notify the sink of the number of flows coming from the transmitter, transmission rate, type of data flow, and needed reliability [21]. After receiving the session initiation packet, the sink records all of the information, adjusts timers for each flow, and sends an acknowledgment.

Whenever sensor nodes receive the ACK, they begin to transport packets to the sink. The session initiation packet is used to organize multiple streams. If the sensor node has multiple sensing devices and needs to transmit data obtained from more than one device, it sends a session initiation packet for all flows. Because the transmission attributes may be mismatched, the source node independently transmits packets belonging to each flow.

In continuous flows, the sink can calculate the expected arrival of the next packet by utilizing the transmission rate. If the sink does not receive a packet within the expected window of time, it then sends an NACK to sensor nodes, notifying them to retransmit the packet.

To resend the lost packets, the transmitted packets are held in the buffer of the sensors. A buffer timer is used to avoid buffer overflow. Here, the size of the buffer is monitored; if it reaches the threshold, the buffer is cleared. The sink cannot measure arrival times of data packets in event-driven flows. Whenever the sink receives a packet, it transmits an ACK to the sender to provide reliability. Provided that the sender does not obtain an ACK in a predetermined time, it resends the packet. Until an ACK is received, the sender will keep the transmitted packets in its buffer.

The session initiation packet is used to indicate the needed reliability for each flow. A running average of the reliability is measured by the sink through the fraction of the packets successfully received for a continuous flow. As long as current reliability fulfills the needed reliability in the session initiation packet, a NACK packet will not send for a lost packet in the flow. For event-driven flows, reliability is defined as the ratio of packets obtained to the maximum sequence numbered packet taken. Before transmitting the packet, a sensor node checks the reliability of transmission by assuming that the packet will get lost. If the obtained reliability satisfies the needed reliability, the node will not buffer the packet, leading to saved memory usage. In STCP, congestion is detected by the intermediate nodes based on queue length. To indicate congestion, a congestion notification bit is set in the header of the STCP packet. Upon getting an STCP packet in a congested network flow, the sink alerts the source to the congestion by setting a bit in the ACK packet. When the source receives this ACK packet, it may change its paths or reduce its transmission rate.

In data-centric applications, because the number of sources may be immense, acknowledgment for reliability will be an extremely resource-consuming task in terms of network resources and energy. Hence, STCP does not employ any ACK-based schemes. It assumes that data from different sensors are redundant, hence, events can be transmitted to the sink in a reliable manner.

Limitations:

- E2E and not scalable: STCP includes E2E congestion-control techniques. Hence, delivering congestion information in a WSN, which includes a huge number of sensor nodes, is a time-consuming process. This process leads to delay in the network.
- Inability to cope with bursty multimedia traffic: Multimedia traffic causes high data rates and thus congestion in the network. Acknowledgment (ACK and NACK) is also a time- and energy-consuming process. The algorithm is weak for high bandwidth applications.

Solutions:

- A smarter algorithm integrating Random Early Detection (RED) and an E2E approach could be tailored to STPC. Multimedia traffic should be differentiated depending on the content of the data. A proactive scheme along with this prioritization scheme could be employed to transmit the packets by taking advantage of the multistreaming scheme of the protocol. Packet loss could be handled separately depending on the reasons of it (congestion or link failure). The techniques for reliability should include a hop-by-hop approach depending on the importance of the packets and reasons for the losses.

2.3 Network Layer Protocols

The network layer employs a routing protocol to find paths between the source and destination [27]. Due to transmission distortions induced by channel problems, energy restrictions, and software or hardware malfunctions, the routing performance may not be acceptable for QoS-based WMSN applications. Thus, the routing protocol is a prominent issue to meet the application layer QoS requirements for transmitting multimedia data. In this section, we discuss several routing protocols and their weaknesses in terms of multimedia transmission. Furthermore, we recommend various solutions to increase the suitability of network protocols for multimedia applications.

2.3.1 Sensor Protocols for Information via Negotiation

The routing protocol, referred to as sensor protocols for information via negotiation (SPIN), is proposed in Ref. [22]. Source adjustment and negotiation mechanisms used by SPIN cover the deficiencies of flooding. Negotiation decreases overlap and implosion problems, while a threshold-based resource-aware algorithm provides the improvement of network lifetime. SPIN employs three different types of messages: Advertisement (ADV), Request (REQ), DATA. A sensor node broadcasts an ADV including metadata, describing the actual data with fewer bytes. Provided that a neighbor decides to receive the data, it transmits a REQ message to the source; in turn, the data are disseminated through the source over the network. In SPIN, each sensor requires knowledge about only its single hop neighbors, hence topological changes are not necessary to handle in the communication process.

Limitations:

- Limited scalability: The SPIN protocol does not adapt well to a WSN size increase. Its performance may decrease dramatically for a larger number of hops.
- Inability to guarantee data delivery: The data may not be transmitted due to uninterested nodes on the path between sender and receiver. As a result, this protocol is not good for applications that need reliability.
- Algorithm overhead: Control messages cause overhead in terms of energy and delay. Generating metadata is also an impractical function for resource-limited nodes in terms of processing, energy, and storage.
- Unbalanced energy consumption: The nodes around the sink could exhaust their energy if there are too many events to be interested by the sink.

Solutions:

- A new algorithm for reliability could be implemented to guarantee data delivery while decreasing message overhead for a small WMSN. By means of the control messages, the algorithm should differentiate multimedia packets depending on importance. Relay nodes receiving data info could apply the same adaptive methods to transmit high priority packets to the destination.

2.3.2 Directed Diffusion

Directed diffusion protocol [20] is convenient in scenarios where the sensor nodes deliver queries for information captured by other nodes, rather than the queries originating only from a sink. By utilizing interest gradients, directed diffusion protocol improves on data diffusion. Wireless sensor nodes identify information amidst the attributes, while the other nodes decide their interest based on these attributes. As long as the sink requires the data to be reported to it, it repeatedly messages its interest. The data are transmitted on the reverse path of the interest diffusion. A gradient built at the time of interest propagation is used for attribution for each path. Negative gradients restrain the distribution of data on a particular path, and positive gradients support data flow along the path.

The advantage of this scheme is that it results in multiple paths from source to sink with different gradients. During transmission, nodes cache or locally transform (aggregate) data in the diffusion model, increasing the scalability of communication and also decreasing the number of message transmissions needed. A reinforcement process is utilized if the sink needs more periodic updates from the sensors, which have discovered an event. Then, the sink broadcasts its interest in a higher data rate requirement. By contrast, if the sink requires only a few updates, it employs negative reinforcement by lowering required data rates. In network processing, the data can be fused into directed diffusion, so that each node actively aggregates queries conveying the same interest and reports including correlated data.

Limitations:

- High complexity: A high degree of complexity results in high latency and extensive usage of computational power and storage.

- High overhead: In-network processing is a drawback for multimedia transport. It requires significant processing power, leading to death of the sensor nodes in the network. As a result, the topology of the network may change and even cause network disruption.
- Weak algorithm for RT and high bandwidth applications: Because directed diffusion is not designed for QoS applications, it does not provide QoS guarantees.

Solutions:

- A new algorithm should be created to utilize multipath technology and leverage in-network processing. A multipath scheme could be used to balance network load and to differentiate packets depending on their importance. When deciding paths, QoS metrics should be considered.

2.3.3 Low-Energy Adaptive Clustering Hierarchy

Low-energy adaptive clustering hierarchy (LEACH) is a well-known hierarchical routing protocol for WSNs. It forms clusters based on signal strength and selects cluster heads (CHs) randomly from the sensor nodes [18]. By means of the rotating CH roles among the sensors, the energy load is distributed to the sensor nodes. The CH nodes aggregate data coming from the sensor nodes in the cluster and then transmit the combined data to the sink. This method decreases the number of messages transported to the sink. MACs based on time division multiple access Time division multiple access (TDMA) and CDMA are utilized to lessen collisions, which may occur inside or outside of the clusters. Data gathering is carried out periodically or on demand in a centralized manner. LEACH is composed of two phases: the setup phase and the steady-state phase. The setup phase includes forming the clusters and electing the CH nodes. In the steady-state phase, data communication travels from the sensors to the sink. In the setup phase, a certain ratio of the nodes (p) is determined as CHs. In this process, a sensor node produces a random number (r). If this number (r) is greater than a predefined value, $T(n)$, the node wins the current round. It is then assigned as a CH.

All chosen CHs transmit an advertisement message to all the non-CH nodes in the WSN. Based on the signal strength of this message, all non-CH nodes determine which clusters they belong to. All non-CH nodes transport a packet to the appropriate CH node so as to join the related cluster. A CH allocates a time slot for each node belonging to it; the nodes in that cluster can only transmit their data to the CH. All nodes are informed of their defined time slot.

The sensors can start sensing and sending data to the CHs in the steady-state phase. The CH node, after receiving all the data, combines it before transporting it to the sink. After a predetermined time, the network triggers the setup phase again and new CHs are selected at the end of this phase. Different CDMA codes are used to decrease interference incoming from the other nodes owned by the other clusters.

Limitations:

- In-network processing issue: In-network processing is a drawback for multimedia transport. It leads to energy consumption by the sensors.
- Higher resource cost: The dynamic nature of the protocol—the CH changes, messaging, etc.—causes additional overhead, which increases energy consumption of the protocol. Additionally, in this protocol, data gathering is performed periodically, which is unsuitable and one of the causes of energy waste for event-based WMSN applications.
- Lack of scalability: LEACH presumes that all nodes have enough power to reach the sink and always have data to transmit. Additionally, it assumes all sensors start with an equal amount of energy in each selection round. Therefore, it may not be appropriate for large networks.

Solutions:

- A new algorithm adjusting channel status and communication load could be implemented with a consideration given to scalability. LEACH should also be improved to account for nonuniform energy nodes.

2.3.4 Power-Efficient Gathering for Sensor Information Systems

Power-efficient gathering for sensor information systems (PEGASIS) [24] is a data-collecting protocol. It is assumed that the topology information is available to all nodes and a leader node can reach the sink in one hop. PEGASIS aims to minimize the distance over which each node sends transmissions, the number of packets that need to be transmitted to the sink, and broadcasting overhead; it also aims to distribute the energy consumption equally across all nodes. Sensor nodes, starting with the node furthest from the sink, build a chain by using a greedy algorithm. The nearest unvisited neighbor is added to the chain at each step. Before beginning data transmission, the chain is established *a priori*; when nodes die out, it is reconstructed. Data fusion or aggregation is performed at every node, so that only one message is forwarded from one node to the next. A node entitled as the leader lastly transports the message to the sink. Leadership is assigned to the next one in a sequential manner and a token is passed on in the chain to transmit the data.

Limitations:

- Delay: Until all of the messages are captured, the leader does not start to communicate with the sink. This situation leads to too much delay for distant nodes in the chain.
- High overhead: PEGASIS has high overhead in terms of delay, computational power, and storage. Aggregation at each hop is a drawback for multimedia transport and may cause erroneous information.
- Lack of scalability: Single-leader concept used in this protocol may be a bottleneck in the network.

Solutions:

- A new scalable algorithm is needed that considers both energy efficiency and delay sensitivity in WMSNs. A smart aggregation algorithm could be performed to decrease the amount of data to be transmitted. This algorithm should decide if aggregation is required or not for the data held in a given sensor node. At this point, sensor nodes can combine their data with the incoming message and then transmit or only behave as relay nodes.

2.3.5 Minimum Energy Communication Network and Small MECN

Location awareness provides an increase in the performance of a network in terms of delay and energy consumption [13]. Minimum energy communication network (MECN) utilizes the location information to make effective routing decisions. It uses a GPS to build and sustain a MECN for a WSN. MECN tries to find a subnetwork with a lower number of nodes where the energy conservation between any two nodes is provided. Global minimum power paths can be found by using a localized search for individual sensor nodes in their transmission region. MECN protocol includes two phases as follows: (1) a sparse graph (enclosure graph) based on two-dimensional positions is constructed. The graph is composed of all the enclosures of each transmitting node. The construction of the graph requires local computations at the nodes. (2) The algorithm finds optimal links in terms of energy waste on the enclosure graph. The distributed Bellman–Ford shortest path algorithm is utilized with energy consumption as the cost metric. Due to MECN self-configuration topology, both node failures and new sensor deployment can be easily handled. Additionally, MECN modifies the minimum cost links adaptively for topological changes.

The small MECN (SMECN) [23] is a modified version of the MECN protocol. MECN assumes that every node can always transport to other sensors, whereas SMECN considers the possible barriers between any two nodes due to availability. MECN and SMECN are categorized into proactive routing protocols, which keep the recent routing information. SMECN is superior to MECN [23] in terms of energy usage and maintenance cost of the links, at the expense of escalated overhead in the algorithm.

Limitations:

- Weak algorithm for RT and high bandwidth applications: Because of its proactive behavior, topological changes cause high delay.
- High overhead: Building a subnetwork with a smaller number of edges leads to overhead.

Solutions:

- A new algorithm should be designed to account for delay factors, leveraging proactive and reactive style. Other QoS parameters as well as energy can be integrated into the MECN during subnetwork building.

2.3.6 Sequential Assignment Routing

The sequential assignment routing (SAR) [35] algorithm is a table-driven multipath protocol that provides QoS-based routing in WSNs. It aims to find optimal routes between the sink and the source in terms of energy usage and reliability. In this protocol, a weighted QoS metric is computed and then minimized so as to multiply the lifetime of the network. SAR uses three considerations: priority levels of the packets, QoS on each path, and energy resources. A multipath routing approach and path restoration techniques are utilized to obviate path failure. A tree is created to obtain multiple paths from the sink to the sensors. During the final part of this procedure, each sensor node is included in more than one path. Node failures because of energy constraints or the wireless environment result in topology changes, hence the path is rebuilt. The sink also recomputes the paths periodically to cope with the changing topology.

Limitations:

- Lack of testing: Performance testing is needed for multimedia transmission.
- Lack of scalability: As the number of nodes in the network is increased, SAR suffers from processing time overhead for handling the tables and status data for each of the sensor nodes.

Solutions:

- SAR does not consider the nature of multimedia data during communication. A new scalable algorithm can be designed to give priority to the multimedia packets in terms of bandwidth utilization and reliability.

2.3.7 A Stateless Protocol for Real-Time Communication (SPEED)

SPEED [17] is a well-known QoS-based routing algorithm that provides real-time anycast for data transmission. Because SPEED is based on hop-by-hop data transmission, it does not need a routing table, which causes minimal control packet overhead. A special packet-naming periodic beacon is used between neighboring nodes. To adapt network condition alterations, two different kinds of beacon packets are used for congestion detection and delay assessment. Using topographic information, data packets are transmitted specifically to the nodes that are nearer to the sink. Among the suitable nodes, the nodes with the minimum estimated delay are selected as intermediate nodes. In the absence of a sensor node to satisfy the delay restraint, the packet is dropped. Although SPEED does not have packet priorities, it provides real-time data communication over WSNs by supplying assurance on the maximal delay.

The Multipath Multi-SPEED Protocol (MMSPEED) protocol [14], which is an enhanced version of the SPEED protocol, supports service differentiation and a probabilistic QoS guarantee. Global network state information and E2E path setup are not required for all functions in MMSPEED. Hence, the protocol is scalable and adaptable to network dynamics. In MMSPEED, different packet transportation speeds are given for different traffic types as stated in their E2E deadlines. The reliability required by the application is provided with a probabilistic multipath forwarding technique that regulates the number of delivery routes in the communication.

Limitations:

- Channel throughput problem: Due to a certain parameter naming the maximum delivery speed, SPEED does not transmit packets at an increased speed, even if the network can present higher rates.
- Trade-off between energy and delay: SPEED and MMSPEED do not consider the energy–delay trade-off. There is no delay guarantee in a dynamically changing network.
- No consideration of Aggregation issues: No network layer aggregation scheme is employed in these protocols. Additionally, MMSPEED does not account for the number of hops between source and destination in making route decisions.
- No consideration of energy issues: SPEED and MMSPEED do not take into account further energy metrics.
- Resource requirement: MMSPEED needs to buffer the reliable forwarding probabilities of the neighboring nodes, which require several updates.

Solutions:

- A modified algorithm should be adopted using a probabilistic approach based on energy constraints. The algorithm could deal with a mapping process from application layer quality to network parameters. It could also be integrated with a localization algorithm to obtain high scalability.

2.3.8 An Energy-Aware QoS Routing Protocol for WSNs

The energy-aware QoS routing protocol for WSNs given in Ref. [3] locates the least costly path by considering energy efficiency and specific E2E delay constraints. The cost of the link is calculated using several transmission metrics such as the energy reserve of the node, energy consumed by transmission, error rate, and so forth. In this protocol, best-effort and real-time traffic are supported by means of a class-based queuing model. The protocol also provides resource distribution for real-time and non-real-time traffic. In case of congestion, a bandwidth ratio r shows the amount of both types of bandwidth-adjusted traffic on certain outgoing links. A number of less costly paths are located and a path is chosen from among those that satisfy the E2E delay requirement. This protocol consistently performs well in terms of QoS and energy parameters. However, an equal r value is utilized for all nodes in the network.

Limitations:

- No prioritization for real-time traffic: The protocol does not contain a function to provide different priorities for multimedia traffic.
- Lack of scalability: Because it calculates multiple paths, for each node the algorithm needs complete information about the network topology.
- Bandwidth sharing issue: The protocol does not support adaptive assigning of bandwidth sharing for different links.

Solution:

- A modified algorithm with scalability could be adopted based on packet-level priority. It should support adaptive bandwidth sharing for different links.

2.4 MAC Layer Protocols

MAC protocols are responsible for channel adjustment and error control–recovery techniques to provide robust, error-free data transfer between the nodes with minimal energy consumption. The quality of wireless links changes dramatically over time for various reasons such as scattering, diffraction, and many more. This dynamic nature of wireless links causes packet losses and also the degradation of perceptual quality of a multimedia application. Hence, new robust transmission mechanisms are required to provide QoS for multimedia [32]. There are two additional QoS metrics that should be considered for multimedia transmission [8,25]: packet latency and multiple priorities for varying services. Due to the high volume of data involved, energy consumption is still an issue for efficient multimedia data transmission for MAC layer protocols.

2.4.1 Sensor MAC Protocol

Sensor MAC (SMAC) focuses on energy efficiency and self-configuration issues to ensure the sustainability of the WSN. SMAC focuses on energy and self-configuration issues to ensure the sustainability of the WSN. It identifies control packet overhead, overhearing, collision, and idle listening as major sources of energy wastage in WSNs. Hence, SMAC utilizes several techniques to decrease energy consumption. To this end, the low duty cycle technique is used for multihop WSNs. To decrease control overhead and to allow traffic-adaptive wake-up, sensor nodes create virtual clusters depending on their sleep periods. The other technique used is channel signaling, which provides a way for nodes to stop overhearing uninteresting traffic. Ultimately, message passing is used to lower contention latency for those WSN applications needing data aggregation. Because of sleep schedules, the energy consumption for idle listening is decreased. The announcements for sleep schedules lead to reducing time synchronization overhead.

Limitations:

- Increased collision probability results from broadcast data packets without RTS/CTS.

- Trade-off between energy and delay/throughput: This protocol preserves energy at the expense of degraded throughput and latency.
- Schedule exchange between neighbors can cause high overhead for video/audio traffic.
- SMAC may not be able to handle synchronization and coordination of the nodes when sensor nodes have a dynamic duty cycle.
- Sudden buffer overflow may occur at the receiver.
- Because sleep and listen periods are predefined, there may be a reduction in the performance of the algorithm for different traffic loads.

Solution:

- An energy-efficient algorithm should be implemented that preserves E2E delay and throughput concerns.

2.4.2 Traffic-Adaptive MAC Protocol

Traffic-adaptive MAC protocol (TRAMA) [30] aims to increase the usage of traditional TDMA in an energy-efficient and collision-free approach. Two techniques are used to reduce energy consumption: (1) assuring unicast and broadcast transmission without collision and (2) assuring that sensor nodes that are not communicating are allowed to enter a low power and idle state. TRAMA partitions time into a series of random-access signaling periods and scheduled-access (transmission) periods. TRAMA chooses one transmitter within each two-hop neighborhood for each time slot. This selection operation dispenses with the hidden terminal problem, and hence the nodes within one hop of the neighborhood of the transmitter are not exposed to collision. TRAMA exhibits higher percentages of sleep time and lower collision probability compared to Carrier sense multiple access (CSMA)-based protocols. Moreover, intended receivers are indicated in a bitmap; less communication overhead occurs for multicast and broadcast traffic compared to other protocols [42].

Limitations:

- High overhead: TRAMA includes a complex election algorithm and data structure and explicit schedule propagation.
- Delay: TRAMA produces a higher queuing delay.

Solution:

- A resource-efficient algorithm could be implemented to prevent delay.

2.4.3 Diff-MAC

Diff-MAC [43] provides differentiated services and hybrid prioritization for WSN applications. It proposes to increase channel utilization with service differentiation and to provide fair and fast distribution of data. Diff-MAC is used for QoS-based WMSN applications, which generally transmit heterogeneous traffic. Several techniques are utilized to provide QoS in WSNs. Video data are fragmented into tiny packets and then transported as a burst that enables a decrease in retransmission cost in case of link failures. The congestion window size is adjusted depending on the data QoS requirements, leading to reduced packet delays and fewer collisions. The protocol adjusts the duty cycle of the sensor nodes by taking into consideration the current traffic class and attempts to balance latency and energy consumption. Finally, fair data distribution among wireless nodes and all traffic classes is provided by intranode and intraqueue prioritization. Diff-MAC is also easily adapted to varying network conditions.

Limitations:

- Due to its network statistics and dynamic adaptation, it has a high complexity. Additionally, Diff-MAC has several complex and overwhelming functions such as monitoring networks statistics and dynamic adaptation.
- Packet latency: Diff-MAC has packet delays due to early sleeping. However, the absence of sleep–listen synchronization among nearby sensor nodes does enhance the scalability of the protocol.

Solutions:

- An energy-delay efficient algorithm should be integrated with this protocol. The algorithm can minimize E2E delay by limiting packet latency. In addition, the complex techniques used in this protocol should be energy-aware.

2.5 Conclusion

WMSN applications have several QoS requirements such as perceptual quality and timeliness. Unstable wireless links and the tight resource constraints of sensor nodes lead to QoS degradation during multimedia communication. Hence, some new approaches are imperative to ensure that the QoS requirements of applications are met. These approaches need to observe the current status of the network and to take action in an adaptive manner to sustain an acceptable level of quality for multimedia transmissions. In this context, providing accurate mapping from application layer perceptual quality requirements to lower layer system parameters is essential. Because application layer requirements depend heavily on the transmitted multimedia data, the characteristics of the data are associated with the communication architecture of WMSNs. To satisfy the QoS-based WMSN applications, the nature of the multimedia data should be integrated with the communication architecture of the sensor networks. In this chapter, we have presented an overview of well-known communication layer protocols from the transport layer to the MAC layer for WSNs. We have highlighted design constraints and open research issues to suggest further avenues of research in the field of QoS provision in WMSNs at the transport, routing, and MAC layers. We believe that this work will lead to reuse of well-known WSN protocols by adapting them for multimedia transmission in WMSNs.

References

1. Aaron, A., Rane, S., Setton, E., Girod, B. Transform-domain Wyner–Ziv codec for video. In: *Proceedings of SPIE Visual Communications and Image Processing*, volume 5308, pp. 520–528. Citeseer, San Jose, CA, 2004.
2. Akan, O.B., Akyildiz, I.F. Event-to-sink reliable transport in wireless sensor networks. *IEEE/ACM Transactions on Networking*, Maryland University, Baltimore, MD, 13(5), 1003–1016, 2005.
3. Akkaya, K., Younis, M. An energy-aware QoS routing protocol for wireless sensor networks. In: *Proceedings, 23rd International Conference on Distributed Computing Systems Workshops, 2003*, pp. 710–715. IEEE, 2003.
4. Akyildiz, I.F., Su, W., Sankarasubramaniam, Y., Cayirci, E. Wireless sensor networks: A survey. *Computer Networks* 38(4), 393–422, 2002.

5. Akyildiz, I.F., Melodia, T., Chowdhury, K.R. A survey on wireless multimedia sensor networks. *Computer Networks* 51(4), 921–960, 2007.
6. Akyildiz, I.F., Melodia, T., Chowdhury, K.R. Wireless multimedia sensor networks: Applications and testbeds. *Proceedings of the IEEE* 96(10), 1588–1605, 2008.
7. Akyildiz, I.F., Melodia, T., Chowdury, K.R. Wireless multimedia sensor networks: A survey. *IEEE Wireless Communications* 14(6), 32–39, 2007.
8. Boluk, P., Baydere, S., Harmanci, A. Robust image transmission over wireless sensor networks. *Mobile Networks and Applications* 16, 149–170, 2011. DOI: 10.1007/s1 1036-010-0282-2.
9. Boluk, P.S., Baydere, S., Harmanci, A.E. Perceptual quality-based image communication service framework for wireless sensor networks. *Wireless Communications and Mobile Computing* 14(1), 1–18, 2011.
10. Boluk, P.S., Baydere, S., Harmanci, A.E. Perceptual quality-based image communication service framework for wireless sensor networks. *Wireless Communications and Mobile Computing* 14(1), 1–18, 2014.
11. Chen, D., Varshney, P.K. QoS support in wireless sensor networks: A survey. In *International Conference on Wireless Networks,* volume 13244, pp. 227–233. Citeseer, Las Vegas, Nevada, 2004.
12. Culurciello, E., Andreou, A.G. CMOS image sensors for sensor networks. *Analog Integrated Circuits and Signal Processing* 49(1), 39–51, 2006.
13. Ehsan, S., Hamdaoui, B. A survey on energy-efficient routing techniques with QoS assurances for wireless multimedia sensor networks. *Communications Surveys & Tutorials, IEEE* 14(2), 265–278, 2012.
14. Felemban, E., Lee, C.G., Ekici, E. MMSPEED: Multipath multispeed protocol for QoS guarantee of reliability and timeliness in wireless sensor networks. *IEEE Transactions on Mobile Computing* 5(6), 738–754, 2006.
15. Feng, W., Kaiser, E., Feng, W.C., Baillif, M.L. Panoptes: Scalable low-power video sensor networking technologies. *ACM Transactions on Multimedia Computing, Communications, and Applications (TOMCCAP)* 1(2), 151–167, 2005.
16. Hao, J., Kim, S.H., Ay, S.A., Zimmermann, R. Energy-efficient mobile video management using smartphones. In: *Proceedings of the 2nd Annual ACM Conference on Multimedia Systems,* pp. 11–22. ACM, Santa Clara, CA, 2011.
17. He, T., Stankovic, J.A., Lu, C., Abdelzaher, T. SPEED: A stateless protocol for real-time communication in sensor networks. *International Conference on Distributed Computing Systems,* 0, 46, 2003.
18. Heinzelman, W.R., Chandrakasan, A., Balakrishnan, H. Energy-efficient communication protocol for wireless microsensor networks. In: *Proceedings of the 33rd Annual Hawaii International Conference on System Sciences, 2000,* pp. 10, IEEE, Island of Maui, 2000.

19. Hyung, S.L., Hee, Y.Y., Jung, H. Context-aware cross-layered multimedia streaming based on variable packet size transmission. In: *Proceedings of the International Conference on Computer Science and Its Applications (ICCSA 2006)*, pp. 691–700. Glasgow, UK, 2006.
20. Intanagonwiwat, C., Govindan, R., Estrin, D. Directed diffusion: A scalable and robust communication paradigm for sensor networks. In: *Proceedings of the 6th Annual International Conference on Mobile Computing and Networking*, pp. 56–67. ACM, Boston, MA, 2000.
21. Iyer, Y.G., Gandham, S., Venkatesan, S. STCP: A generic transport layer protocol for wireless sensor networks. In: *Proceedings, 14th International Conference on Computer Communications and Networks, 2005. ICCCN 2005*, pp. 449–454. IEEE, San Diego, California, 2005.
22. Kulik, J., Heinzelman, W., Balakrishnan, H. Negotiation-based protocols for disseminating information in wireless sensor networks. *Wireless Networks* 8(2), 169–185, 2002.
23. Li, L., Halpern, J.Y. Minimum-energy mobile wireless networks revisited. In: *IEEE International Conference on Communications, 2001. ICC 2001*, volume 1, pp. 278–283. IEEE, Helsinki, Finland, 2001.
24. Lindsey, S., Raghavendra, C.S. PEGASIS: Power-efficient gathering in sensor information systems. In: *IEEE Aerospace Conference Proceedings, 2002*, volume 3, pp. 3–1125. IEEE, Big Sky, Montana, 2002.
25. Misra, S., Reisslein, M., Xue, G. A survey of multimedia streaming in wireless sensor networks. *IEEE Communications Surveys & Tutorials* 10(4), 18–39, 2008.
26. Nachman, L., Huang, J., Shahabdeen, J., Adler, R., Kling, R. Imote2: Serious computation at the edge. In: *International Wireless Communications and Mobile Computing Conference, 2008. IWCMC'08*, pp. 1118–1123. IEEE, Crete Island, Greece, 2008.
27. Pantazis, N.A., Nikolidakis, S.A., Vergados, D.D. Energy-efficient routing protocols in wireless sensor networks: A survey. *Communications Surveys & Tutorials, IEEE* 15(2), 551–591, 2013.
28. Polastre, J., Szewczyk, R., Culler, D. Telos: Enabling ultra-low power wireless research. In: *4th International Symposium on Information Processing in Sensor Networks, 2005. IPSN 2005*, pp. 364–369. IEEE, UCLA, Los Angeles, California, 2005.
29. Rahimi, M., Baer, R., Iroezi, O.I., Garcia, J.C., Warrior, J., Estrin, D., Srivastava, M. Cyclops: In situ image sensing and interpretation in wireless sensor networks. In: *Proceedings of the 3rd International Conference on Embedded Networked Sensor Systems*, p. 204. ACM, San Diego, CA, 2005.
30. Rajendran, V., Obraczka, K., Garcia-Luna-Aceves, J.J. Energy-efficient, collision-free medium access control for wireless sensor networks. *Wireless Networks* 12(1), 63–78, 2006.
31. Rowe, A., Goode, A., Goel, D., Nourbakhsh, I. CMUcam3: an open programmable embedded vision sensor. Technical Report CMU-RI-TR-07-13, Robotics Institute, Carnegie Mellon University, Pittsburgh, PA, 2007.

32. Sarisaray-Boluk, P. Performance comparisons of the image quality evaluation techniques in wireless multimedia sensor networks. *Wireless Networks* 19(4), 443–460, 2013.
33. Sen, J., Bhattacharya, S. A survey on cross-layer design frameworks for multimedia applications over wireless networks. *International Journal of Computer Science and Information Technology (IJCSIT)* 1(1), 29–42, 2010.
34. Shu, L., Zhang, Y., Yu, Z., Yang, L.T., Hauswirth, M., Xiong, N. Context-aware cross-layer optimized video streaming in wireless multimedia sensor networks. *The Journal of Supercomputing* 54(1), 94–121, 2010.
35. Sohrabi, K., Gao, J., Ailawadhi, V., Pottie, G.J. Protocols for self-organization of a wireless sensor network. *IEEE Personal Communications* 7(5), 16–27, 2000.
36. Soro, S., Heinzelman, W. A survey of visual sensor networks. *Advances in Multimedia*, 1–21, 2009.
37. Su, W., Cayirci, E., Akan, O. Overview of communication protocols for sensor networks. In: *Handbook of Sensor Networks: Compact Wireless and Wired Sensing Systems* (Eds. M. Ilyas, I. Mahgoub), pp. 374–392, Boca Raton, FL, USA, CRC Press, p. 374, 2004.
38. Wan, C.Y., Campbell, A.T., Krishnamurthy, L. PSFQ: A reliable transport protocol for wireless sensor networks. In: *Proceedings of the 1st ACM International Workshop on Wireless Sensor Networks and Applications*, pp. 1–11. ACM, Atlanta, GA, 2002.
39. Wan, C.Y., Campbell, A.T., Krishnamurthy, L. Reliable transport for sensor networks. In: *Wireless Sensor Networks* (Eds. C.S. Raghavendra, K.M. Sivalingam, and T. Znati), pp. 153–182. Springer, Berlin, Germany, 2004.
40. Wan, C.Y., Eisenman, S.B., Campbell, A.T. CODA: Congestion detection and avoidance in sensor networks. In: *Proceedings of the 1st International Conference on Embedded Networked Sensor Systems*, pp. 266–279. ACM, Los Angeles, CA, 2003.
41. Wang, C., Sohraby, K., Lawrence, V., Li, B., Hu, Y. Priority-based congestion control in wireless sensor networks. In: *IEEE International Conference on Sensor Networks, Ubiquitous, and Trustworthy Computing, 2006*, volume 1, pp. 8. IEEE, Newport Beach, CA, 2006.
42. Yahya, B., Ben-Othman, J. Towards a classification of energy aware MAC protocols for wireless sensor networks. *Wireless Communications and Mobile Computing* 9(12), 1572–1607, 2009.
43. Yigitel, M.A., Incel, O.D., Ersoy, C. QoS-aware MAC protocols for wireless sensor networks: A survey. *Computer Networks* 55(8), 1982–2004, 2011.
44. Zacharias, S., Newe, T. Technologies and architectures for multimedia-support in wireless sensor network. In: *Smart Wireless Sensor Networks*, (Ed. Y. K. Tan) InTech, Rijeka, Croatia, pp. 18–25, 2010.

3

Multimedia Communication in Cognitive Radio Ad Hoc and Sensor Networks

MUSTAFA OZGER, ECEHAN B. PEHLIVANOGLU, AND OZGUR B. AKAN

Contents

3.1 Introduction

Small and low-cost sensor nodes are available, thanks to developments in micro-electro-mechanical systems (MEMS) technology. These sensor nodes have irreplaceable batteries and they are deployed in specific regions of interest. The deployment of these sensor nodes forms wireless ad hoc networks, namely wireless sensor networks (WSNs). The application areas of these networks are environmental or habitat monitoring, military surveillance, medical applications, multimedia applications, and so forth. The sensor nodes sense the environment—heat, pressure, sound, light, or motion depending on the application—and form packets related to the observations. The sensor nodes collaborate with each other to convey these packets in a multihop manner to a base station or sink [1].

Reliable and timely delivery of data packets is a vital requirement for multimedia communication. The inherent demands of multimedia communication are limited delay, large bandwidth, jitter control, no abrupt changes of transmission rate, and loose reliability [2]. Hence, a new networking paradigm—wireless multimedia sensor networks (WMSNs)—has been proposed to realize multimedia communication in WSNs. Multimedia communication demands more resources than mainstream data-sensing applications. Event features in sensor networks and the packets generated by source nodes in ad hoc networks contain large byte streams, because they are in the form of multimedia, for example, video, audio, and still images. Hence, the main challenges posed by multimedia communication are high bandwidth demand and strict time constraints [3].

Ever-increasing demand in wireless communication has led to the spectrum scarcity problem. Furthermore, the electromagnetic spectrum is not utilized efficiently due to the fixed frequency assignment approach. Cognitive radio (CR) technology has become the solution to the problems of spectrum scarcity and inefficient utilization. Communication between wireless nodes becomes spectrum-aware by cognitive cycle operations, namely spectrum sensing, spectrum decision, and spectrum handoff [4]. The inherent features of CR have increased electromagnetic spectrum efficiency and have overcome the spectrum scarcity challenge. The main CR application areas are ad hoc networks and sensor networks [5].

A typical CR ad hoc network consists of two types of users: primary users (PUs) and secondary users (SUs). PUs are the legacy users of the licensed spectrum. SUs are the CRs, and they utilize the licensed bands opportunistically [6]. The CR capability of wireless nodes makes them adaptable to environmental changes in the network. A CR node can find vacant spectrum bands by spectrum sensing and can make use of them through the spectrum decision process. If a channel that is being utilized becomes occupied by a PU, the communication continues on another vacant channel through the spectrum handoff function. Hence, CR can change its operating parameters to enable seamless communication.

WSNs are characterized by a fixed frequency assignment policy; hence, they also suffer from the spectrum scarcity challenge [5]. The event-driven nature of WSN communications causes bursty traffic. The injection of event packets leads to packet collisions and excessive contention delay due to high traffic and high node density in the network. Furthermore, sensor nodes have inherent limitations in power, communication, processing, and memory resources. Hence, the CR sensor network (CRSN) has been proposed by integrating CR capability into sensor networks. Providing sensor nodes with CR capability increases the efficiency of overall spectrum utilization and decreases the probability of collision and contention by utilizing multiple channels opportunistically. Adaptability to the existing spectrum meets the unique requirements of WSNs. To this end, CRSNs benefit from the potential advantages of the salient features of dynamic spectrum access, such as adaptability to environmental changes to decrease collision and to reduce power consumption and accessing multiple channels to provide flexibility in spectrum usage.

Sensor nodes in CRSNs sense the environment as well as the spectrum. The event readings of the sensor nodes in the event region are collaboratively conveyed in a multihop manner through vacant channels. A typical CRSN architecture for multimedia communication is shown in Figure 3.1. In this architecture, multimedia CRSN nodes and PUs coexist and there is a sink with CR capability. Base stations serve as PUs, and CRSN nodes communicate without interfering with PUs. The CR-capable sink can receive multiple data with its CR-capable transceivers. Furthermore, a typical CRSN topology can be ad hoc, clustered, hierarchical, and mobile, depending on the application.

A wide range of potential applications for CRSNs exists in the literature. These applications include indoor sensing applications, multiclass heterogeneous sensing applications, and real-time surveillance

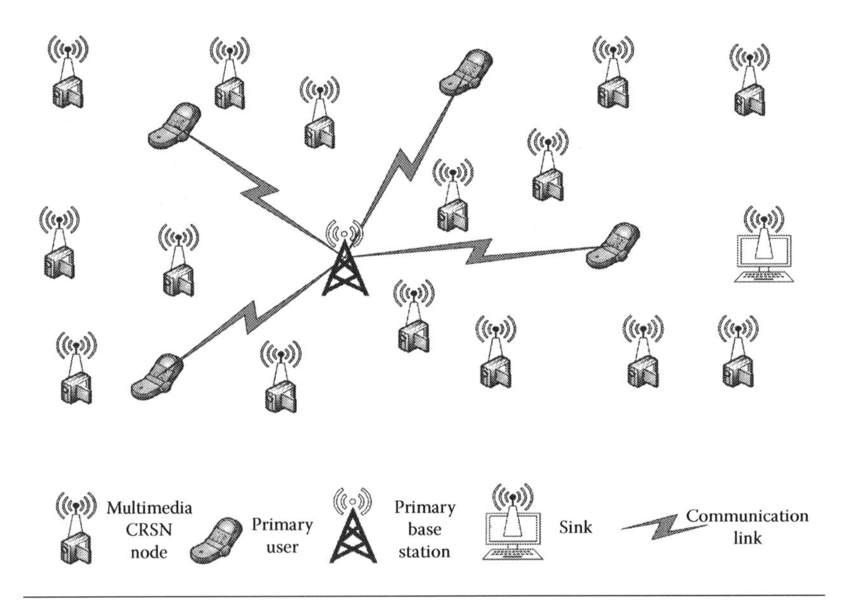

Figure 3.1 A typical cognitive radio sensor network architecture.

applications [5]. Furthermore, one of the most significant areas of application is multimedia communication [3].

Multimedia communication has been investigated in a wireless sensor and ad hoc network context extensively [2,3,7]. However, there is no research on multimedia communication in CR ad hoc and sensor networks. The dynamic radio environment poses distinctive challenges in multimedia communication. Multimedia communication requires seamless communication, bounded delay, and a certain level of quality of service (QoS). However, with CR capability, communication may be interrupted by the arrival of PUs, resulting in delay, route changes, and variation in QoS. Hence, multimedia communication in a CR network (CRN) regime intensifies existing challenges. In this chapter, we survey the approaches proposed for each network layer in WSNs, CRNs, and WMSNs; we examine the challenges and potential approaches to overcome them.

We survey existing studies and present research opportunities on multimedia communication from the perspective of network layers in CR ad hoc and sensor networks. We examine the state-of-the-art approaches related to WMSN communication and how these methods can be applied to multimedia communication in CR ad hoc and sensor networks. We clearly indicate the challenges posed by CR and sensor networks for multimedia communication and investigate how the union of sensor networks and CR in multimedia communication poses new challenges.

Multimedia streams must be conveyed in a timely manner. However, CR introduces some delays due to cognitive cycle operations and coordination activities, such as the need for communicating parties to tune to the same channel. Hence, these are the specific challenges caused by enabling wireless nodes with CR capability. Furthermore, multimedia applications require limited delay and mandate the transport of packets in a timely manner. In the light of the above-mentioned challenges and limitations, we investigate the existing protocols and open issues. In the application layer, we investigate multimedia encoding techniques and present open issues for CR sensor and ad hoc networks. We examined the existing transport layer approaches for multimedia communication in WSNs, CRNs, and CRSNs, and we present the potential approaches for CR and ad hoc networks according to reliability of the transport and congestion control. We investigated routing issues in the network layer by examining the previous works on WSNs, CRNs, and CRSNs. We present potential approaches in the medium access control (MAC) layer for medium access control protocols for CR ad hoc and sensor networks. We outline the open research issues in each layer to facilitate multimedia communication in CR ad hoc and sensor networks.

3.2 Factors Influencing Multimedia Communication in Cognitive Radio Ad Hoc and Sensor Networks

Designing protocols for CR ad hoc and sensor networks imposes limitations such as constrained energy and memory capability, decentralized architecture, node failures, dynamic radio environment, and so on. Energy and memory limitations are inherited from the WSN paradigm; opportunistic spectrum access is inherited from the CR capability. These challenges are amplified by the union of the sensor network paradigm and CR. On the other hand, multimedia communication has unique requirements—high bandwidth demand, application-specific QoS requirements, and delay bound [3]. These requirements must be revisited from the perspective of the dynamic radio environment. Satisfying these requirements is the main motivation to facilitate multimedia communication. These requirements are also the main factors influencing the quality of multimedia communication. These factors are explained as follows.

- High bandwidth demand: Multimedia traffic contains large packets and they must be transported to the destination within a certain delay. Hence, high bandwidth is a vital requirement for real-time multimedia communication. However, the dynamic radio environment poses a significant challenge here because of PU activities. The channel conditions change with the arrival of PUs and spectrum handoff.
- QoS requirements and limited delay: Either processing time or communication latency in WMSN may cause delay [3]. Furthermore, channel access delay should be taken into account for the delay in CR ad hoc and sensor networks. Low delay and high throughput are the main QoS requirements in multimedia communication. In addition, jitter degrades the performance of

continuous media such as audio and video. It can be avoided by replay buffering at the receiver side.

- Dynamic radio environment: The cognitive capability of nodes offers flexibility in utilizing communication channels. However, they pose significant challenges. For example, the coordination of communication is a difficult task, since the communication between cognitive nodes requires tuning to the same channel for communication.
- Power consumption: Nodes have irreplaceable batteries; hence, energy efficiency is a vital requirement. Collision, unnecessary packet transmission, extensive overhead due to communication coordination, and high duty cycle operation are the most important factors in power consumption.

3.3 The Application Layer in Multimedia Communication for Cognitive Radio Ad Hoc and Sensor Networks

The application layer is perhaps the most important platform in multimedia communication, since it governs the communication quality of the end user. In a WSN or CRSN, the application layer needs to compress the field data efficiently and obtain a proper representation to it, such that data are transported via wireless links in multihop manner.

The application layer is responsible for

- Managing traffic based on application requirements
- Performing source coding via a suitable approach based on the resource constraints of the sensor nodes

Both of these challenging tasks were previously investigated in the context of WSNs. However, intermittent communications in the CR concept and its additional operational requirements necessitate revisiting these tasks and assessing their applicability to CRSNs.

3.3.1 Challenges and Requirements

The challenges faced by the application layer in WMSNs and multimedia communication in CRSNs are mostly driven by the high bandwidth multimedia nature of the data to be communicated. We invite users interested in general challenges on application layer of traditional WSN to read Ref. [1], an excellent survey of WSNs and their challenges. Refs. [2,3,7] are excellent sources for a general overview on the challenges and requirements of the application layer for WMSNs. Based on the tasks of managing traffic with respect to the application requirements and source coding, the challenges of the application layer in WMSNs and CRSNs will be investigated along different dimensions, as shown in Figure 3.2.

- QoS is an important metric to be considered in sensor networks, even more in multimedia sensor networks. In that sense, based on application requirements, sensor nodes may need to differentiate different types of traffic. Therefore, existing measures in this respect and their applicability to CRSNs are worthy of consideration.
- Multimedia applications demand high bandwidth, which is scarce in the industrial, scientific, and medical bands populated by WSNs. Although CRSNs can offer a remedy to the bandwidth problem, efficient source coding techniques are needed to carry out source-to-sink multimedia relaying in an energy-efficient manner. Depending on the network and application, distributed source coding (DSC) or individual source coding may be used in WMSNs [2]. Moreover, different compression techniques exhibit differences in resilience to errors, energy efficiency, and adaptability to network conditions, among other properties. Communication is intermittent in CRSNs, with constant adaptation of communication parameters and additional tasks regarding cognitive cycle bringing additional challenges in communications and energy expenditure. Therefore, compression techniques need to be revisited for multimedia communications in CRSNs.

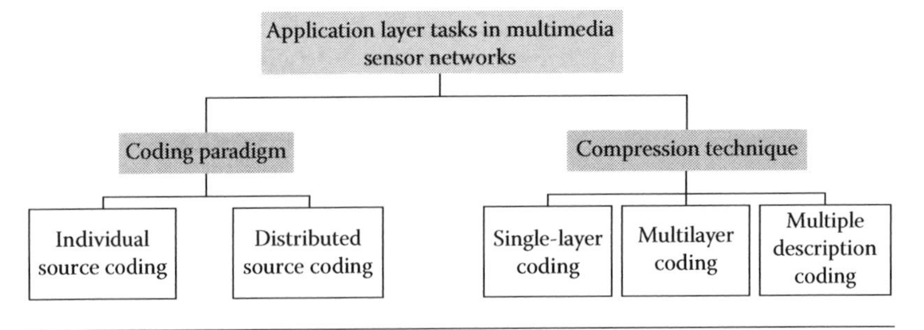

Figure 3.2 Application layer tasks in a multimedia sensor network.

3.3.2 Existing Application Layer Solutions

3.3.2.1 WMSNs In traffic management, application layer responsibility in a WMSN depends on whether the application is delay-tolerant and is concerned about a multimedia stream or a specific data. In real-time applications, tolerance for delay is very low, which in turn requires CR nodes (for CRSNs) to carry out the cognitive cycle in a quicker manner, with the shortest possible spectrum-sensing duration and acceptable false alarm and false detection levels. In real-time applications, greedy forwarding and nearest neighbor forwarding should be assessed to determine which would cause less delay, and action should be taken, which in turn requires cross-layer coordination. In a delay-tolerant setting, CRSN nodes can periodically turn on and off their transceivers, field-sensing units, or both. To date, no study has considered a sleeping schedule for transceivers and field-sensing units of CRSN nodes at the same time, which would be governed by the traffic-management task of the application layer.

In terms of the coding paradigm, with individual source coding, each node codes its own field readings independently [2]. Simplicity is an advantage of individual source coding because it does not require communication and/or coordination between nodes for coding purposes. Such individual coding has been shown to be suboptimal, especially if the spatiotemporal correlation of the sensed field is significant [8]. Accordingly, in a WMSN, it is possible to choose a representative number of nodes M for an event and adjust their reporting frequency f such that significant power savings can be achieved while conforming to a maximum distortion at the sink, that is, $D(f, M) < D_{max}$.

The video compression technology H.264 possesses excellent rate-distortion performance with its advanced compression techniques [9]. The newly proposed H.265 [10] will even surpass it in terms of compression efficiency. Nevertheless, such predictive coding schemes bring processing complexity and require excessive energy consumption for compression. In addition, predictive coding schemes require complex encoders and decoders. It is not feasible for resource-constrained sensor nodes to use these complex encoders. To solve these problems, DSC has been proposed as the paradigm to balance communication needs and processing cost with a good rate-distortion performance.

Slepian and Wolf [11] first proved that, for lossless compression of two correlated signals, separate simple encoding and joint decoding is as efficient as joint decoding. Considering X and Y as two correlated random variables, from Shannon's source coding theorem, a rate $R \geq H(X, Y)$ is sufficient. This may be obtained by a rate $R_Y \geq H(Y)$ and $R_X \geq H(X \mid Y)$, which in turn produces a total rate of $R_X + R_Y \geq H(X \mid Y) + H(Y) = H(X, Y)$. Slepian and Wolf stated that $R \geq H(X, Y)$ is sufficient for the separate encoding of correlated X and Y, which translates to $R_X \geq H(X \mid Y)$ being achievable without knowing the explicit Y at the encoder but knowing the joint statistics of X and YY only. Wyner–Ziv coding was proposed as an extension to Slepian and Wolf, where correlated discrete signals are coded in a lossy manner this time, with respect to a fidelity criterion. Wyner–Ziv is challenged, though, by the fact that obtaining explicit joint probability density functions to compress the sources (nodes) in a distributed manner is difficult [12]. Exploitation of spatial correlation among sources is the key to overcoming this challenge. This is natural, given the fact that video sources are usually anisotropic. From a CRSN perspective, leveraging spatial correlation via DSC is still more challenging, since DSC would require synchronization between two nodes to be jointly decoded at the sink. Due to temporal variation of licensed user traffic, even neighboring nodes might not be able to communicate for proper synchronization, especially in delay-intolerant applications.

DSC via the Wyner–Ziv paradigm (simple encoder at nodes, complex decoder at sink) is a very attractive technique for exploiting the temporal correlation of each source's readings within itself. DSC is indeed one of the hot topics in research, with various studies conducted to find a balance between computation and communication efficiency in sensor networks [13,14,15,16]. On exploitation of temporal correlation of each node, routing schemes may need to be revisited particularly in the case of CRSNs, where routes to the sink may lose connectivity, even permanently, depending on the PU behavior.

On the compression technique side, single layer and multilayer coding techniques have been nicely summarized in Ref. [2].

Accordingly, in single layer coding, a reference layer and only consecutive differences between frames are transmitted until the next reference frame. This type of coding does not provide error resilience. Nevertheless, in underlay cognitive networks, besides bad channel conditions for SUs, a frame is to be retransmitted (potentially through another channel) regardless once it is interrupted via licensed user traffic. However, for CRNs using underlay spectrum sharing, transmission can still continue until the tolerable threshold limit on licensed users has been crossed by the SUs. In that sense, it is more suitable to employ multilayer coding, which allows for joint source and channel coding, and hence error resilience, in underlay CRSNs.

3.3.2.2 Cognitive Radio and Sensor Networks There are far fewer studies regarding multimedia communications in CRNs and CRSNs than in WSNs, given that they are relatively new paradigms.

Ref. [17] is a related work in the context of CRNs, where QoS is optimized by varying the intra-refreshing rate together with the access strategy and spectrum sensing of CR nodes. The idea with the intra-refreshing rate is that an intra-coded macroblock does not depend on previous blocks, which might have been corrupted by errors in previous transmission, thus limiting error propagation and lowering of QoS. Nevertheless, because the study focuses on CRNs, the assumed multimedia model is based on advanced compression techniques of H.264 and alike, which are not suitable for the low complexity nodes of CRSNs.

Dastpak et al. [18] also study multimedia communications in CRN. A streaming optimization problem is set and is solved based on the bandwidth budget available to CR nodes. The results reveal that group-of-picture (GoP)-based optimization is less computationally expensive, but it also has slightly less QoS compared to frame-based optimization. Again, the work is based on the H.264 scalable video codec and is not directly applicable to resource-constrained CRSNs.

Ref. [19] is an interesting work in CRSNs regarding the optimal allocation of power to two sensing tasks, namely field sensing and spectrum sensing. Although the results are insightful, the network model needs to be extended to a multihop network, which is usually the case in CRSNs.

3.3.3 Open Research Issues for the Application Layer

Several open issues exist in CRSN regarding the application layer, as follows:

- For traffic management, an application layer protocol that decides upon sleep and wake-up schedules of the transceivers and field sensor of CRSN nodes is needed for enhanced network lifetime.
- An energy-efficient, cross-layer scheme for source and channel coding in CRSN is needed to ensure reliable monitoring.
- Routing schemes need to be revisited in the CRSN aspect, to make sure that Wyner–Ziv coding that is targeted to leverage node-based temporal correlation makes each packet arrive at the sink, given that connectivity in CRSNs is intermittent among many nodes.

3.4 The Transport Layer in Multimedia Communication for Cognitive Radio Ad Hoc and Sensor Networks

The transport layer provides end-to-end reliability for data transfer and congestion and flow control in traditional computer networking. Transport layer protocols regulate data flows, control congestion, decrease packet loss, and guarantee a certain level of end-to-end reliability in WSNs [20]. However, due to the inherent features of WSNs, such as network topology, traffic characteristics, and their application-dependent nature, the traditional transport layer protocols such as transmission control protocol (TCP) and user datagram protocol (UDP) do not suit WSN.

Transport layer solutions are vital for the realization of multimedia communication in CR ad hoc and sensor networks. Reliable data delivery of event features and the amount of traffic injection upon event detection are the main issues of the transport layer. The existing challenges for CR and sensor networks need to be reconsidered for the realization of the multimedia communication in CR ad hoc and sensor networks. Apart from the WSN paradigm, CR ad hoc and sensor networks impose new challenges for transport layer functionalities. To achieve efficient transport layer solutions, the unique characteristics of CR ad hoc and sensor networks paradigm and the distinctive requirements of multimedia communications must be analyzed in detail to develop transport layer solutions.

3.4.1 Challenges and Requirements

The main duties of the transport layer are reliability and congestion control. Multimedia communication in wireless networks is delay-sensitive data communication. Hence, timely delivery of data packets is more important for performance than reliability. Delay bound and strict QoS demands impose significant challenges on communication. The multimedia packets have a higher size than conventional data packets. Hence, the potential delays cause buffer overflows and the loss of data packets.

In sensor networks, event readings are injected to the network, and this injection creates high and bursty traffic. This situation requires a certain level of congestion control and end-to-end reliability to preserve the limited resources of the nodes. The end-to-end reliability is achieved by the arrival of a sufficient number of event packets at the sink. Congestion control is enabling the event traffic toward the sink to not exceed the capacity of ad hoc or sensor networks. These functionalities of the transport layer are used to provide energy efficiency and to prevent the waste of communication resources [5].

CR has unique features that should be carefully considered to propose efficient solutions. The inherent features of CR give rise to unique challenges, which can be itemized as follows.

- Spectrum sensing: Spectrum sensing is a vital cognitive cycle operation that provides information about spectrum availability. During spectrum sensing, the transmission function is paused, since they cannot be performed simultaneously. The operating channels can be changed after channel sensing. The channel characteristics may change or it may even become occupied by a PU. Furthermore, the channel-sensing period may be changed due to spectrum efficiency and interference avoidance [21]. The scheduling of sensing duration must be organized to decrease collision probability. Different sensing periods of nodes results in variation of packet latency, that is, jitter. Transport layer functionality rate control must consider this heterogeneity in the spectrum-sensing duration that causes the degradation of transport layer performance.

- Interference by the PU: The main spectrum-sensing technique is energy detection. This technique may cause faulty detection; thus, the transmission continues despite the existence of a legacy user on that communication band. Furthermore, a packet transmitted by a CR source node may be received erroneously. The precious resources of wireless nodes are wasted. The probability of successful packet transmission decreases with PU activities.
- Spectrum handoff: During communication between event and sink in CRSNs, and between source and destination in cognitive radio ad hoc networks (CRAHNs), changes in licensed user activity are observable. These changes may degrade the performance of the secondary network. Upon detection of a licensed user in the communication channel, spectrum handoff is performed. The communication is stopped until a new vacant channel can be found. In addition, the delay characteristics and the loss rate change on the communication link, which affects the communication performance.
- Dynamic spectrum access: If a CR does not find a vacant band, then it has to wait for a spectrum opportunity. This waiting time causes extra time delay for packet transmission. In sensor networks, the data packet may become obsolete due to the delay.
- Spectrum mobility: The CR capability of CR ad hoc and sensor nodes provides a change of operating parameters in case of spectrum handoff. The channel and delay characteristics, and also the level of noise change due to the variations in the operating channel parameters.
- Spectrum coordination: The dynamic spectrum environment requires spectrum coordination. The exchange of control packets for channel selection and sensing duration are the main functionalities of spectrum coordination. Spectrum coordination is highly necessary to perform cognitive cycle operations such as spectrum decision and spectrum handoff.

The challenges posed by the CR ad hoc and sensor network paradigm are amplified by the unique requirements of multimedia communication. The dynamic spectrum access scheme provides gain by opportunistic channel access; however, licensed user activities may deteriorate multimedia communication. Furthermore, energy efficiency is an important challenge for resource-constrained wireless nodes. Hence, dynamic spectrum access (DSA) and licensed user activities should be carefully examined to avoid unnecessary power consumption due to collision, interference, and loss through buffer overflow.

3.4.2 Existing Transport Layer Solutions

3.4.2.1 WSNs The main duties of transport protocols are to mitigate congestion and packet losses and to fairly allocate limited bandwidth between the wireless nodes. Hence, we propose a number of WSN transport protocols, according to energy efficiency, reliability, and fairness metrics.

Different techniques can be combined to control and mitigate congestion in WSNs. In Ref. [22], three techniques—hop-by-hop flow control, a source rate limiting scheme, and a prioritized MAC layer that gives priority to backlogged nodes—are combined to mitigate and control congestion. Congestion is detected when the fraction of space of the output queue is less than a certain threshold [22]. To mitigate congestion, hop-by-hop rate adjustment is performed. By contrast, Wan et al. [23] detect congestion in WSNs by also observing the state of the channel for a certain period of time. If channel utilization surpasses a certain threshold, the congestion bit in the outgoing packet is set to one. Congestion mitigation is performed by hop-by-hop back pressure to locally change congestion policy or end-to-end source rate regulation.

Congestion can be detected in several ways in WSN. For example, in Ref. [24], the authors look at packet service time to find congestion. On the other hand, Ref. [20] uses interarrival times of the packets in addition to the service time of the packet to detect congestion. Furthermore, congestion is mitigated by hop-by-hop rate adjustment in Refs. [20,24]. Table 3.1 outlines congestion detection and mitigation techniques used in the literature.

Table 3.1 Existing Congestion Control Protocols in Wireless Sensor Networks

PROTOCOL	CONGESTION DETECTION	CONGESTION MITIGATION
Fusion [22]	Observing output queues	Hop-by-hop rate adjustment
CODA [23]	Observing state of channel	Hop-by-hop rate backpressure
Congestion Control & Fairness Protocol [24]	Observing service time of packets	Hop-by-hop rate adjustment
PCCP [20]	Observing inter-arrival time of packets and service time	Hop-by-hop rate adjustment

Multimedia communication in CR ad hoc and sensor networks requires efficient congestion detection and mitigation mechanisms, because congestion strongly affects real-time communication. Large packet size, the dynamic radio environment, and limited capabilities of sensor nodes necessitate novel schemes by considering the challenges posed by the CRSN paradigm.

Apart from congestion detection and mitigation techniques, reliability is another important functionality of the transport layer in WSNs. Reliable multi-segment transport (RMST) [25] provides end-to-end reliability for all the packets in WSN and does not take into account the correlated nature of the event packets. Packet reliability is provided by requesting data by NACK from the source or the cache nodes. Multi-path and multi-SPEED (MMSPEED) [26] provides reliability at different QoS levels, in addition to timeliness. It utilizes multipath method to increase the delivery probability of packets and to propose a dynamic compensation scheme such that inaccurate decisions made by a node in the path can be compensated for along the route to the sink. Pump slowly fetch quickly (PSFQ) [27] injects or pumps messages slowly to the network. Furthermore, the intermediate nodes cache data for local loss recovery. If there is a gap in the received file fragments, a NACK is sent in the reverse path to recover the lost packets. However, packet loss due to congestion is not addressed in PSFQ.

The transport layer requires end-to-end reliability in wireless ad hoc networks. However, sensor observations are correlated in sensor networks [8]. Hence, end-to-end reliability is not a strict requirement. A sufficient number of sensor observations must be collected by the sink to extract event features with a certain estimation distortion. This reliability model is known as *event reliability* [28]. TCP protocols, which require retransmission of lost packets, are not efficient for the WSN paradigm because retransmission consumes the scarce resources of sensor nodes.

End-to-end reliability is provided at the sink or source nodes. In Ref. [28], the application-dependent reliability level is satisfied at the sink according to the reporting rate of source nodes in the event region. If the packets received according to sensor reports satisfy the reliability level, the reporting rate is decreased. Otherwise, the reporting rate is increased to achieve the desired level of event reliability. With this adaptive feature of event-to-sink reliable transport (ESRT), the event is extracted as a result of the correlated nature of sensor observations.

Rate-controlled reliable transport (RCRT) [29] is a reliable transport protocol for applications that are not tolerant to packet losses. It utilizes end-to-end loss recovery. Furthermore, congestion detection and rate adaptation are done at the sink. It implements a NACK-based end-to-end loss recovery scheme to guarantee reliability. It also utilizes cumulative ACK to remove the packets at the source node that are received by the sink.

Wireless sensor and actor networks are heterogeneous networks where sensor nodes and actors coexist. Sensors observe the environment and actors perform actions according to the sensor observations. $(RT)^2$ [30] proposes real-time reliable transport for sensor–actor and actor–actor communication. It considers event reliability as in ESRT [28].

In CR ad hoc and sensor networks, reliable delivery of all multimedia communication data packets is not necessary due to the correlated nature of sensor observations. Hence, the protocols must consider event reliability and offer a certain level of QoS. Table 3.2 outlines reliable transport protocols in WSNs and their advantages and disadvantages for multimedia communications.

3.4.2.2 Cognitive Radio Ad Hoc and Sensor Networks The unique constraints encountered at the transport layer in WSNs are amplified by the CR capability in CRAHNs. Cognitive cycle operations affect the transport layer protocol design. For example, transport layer functionality is disrupted by spectrum sensing and, if a timeout occurs, the source rate is decreased without congestion. The delay between two nodes increases by spectrum switching due to PU arrival on the communication channel. In Ref. [31], the optimal balance is achieved between the spectrum-sensing duration and the throughput

Table 3.2 Reliability of Transport Protocols in Wireless Sensor Networks

PROTOCOL	MULTIMEDIA COMMUNICATION ISSUES
RMST [25]	– End-to-end reliability for all packets
	– No consideration of correlation between packets
	– NACK-based retransmission
MMSPEED [26]	+ Reliability at different quality of service levels and timeliness
	+ Multipath support
	+ Dynamic compensation against inaccurate decisions
PSFQ [27]	+ Caches data for loss recovery
	– NACK-based retransmission
	– No congestion loss consideration
ESRT [28]	+ Event reliability
	+ Reporting rate adjustment
RCRT [29]	+ Real-time support
	– End-to-end loss recovery by NACK
$(RT)^2$ [30]	+ Event reliability
	+ Real-time reliable transport

Note: Advantages and disadvantages are indicated by + and –, respectively.

of the CRN. It is known that if sensing duration increases, end-to-end throughput decreases [32]. Spectrum switching due to PU activities causes changes in channel characteristics. These changes result in variation of the bandwidth and hence the TCP congestion window is adaptively modified [31]. Consequently, classical TCP is adapted to a dynamic radio environment. In an earlier work, transport protocol for cognitive radio ad hoc networks (TP-CRAHN) [33] was proposed for establishing a spectrum-aware window-based transport layer protocol.

Cross-layer design is also used to improve the performance of the transport layer in CRNs. In Ref. [34], the cognitive cycle operations, physical layer coding and modulation scheme, and data-link layer frame size are concurrently optimized to maximize TCP throughput in CRNs. Furthermore, the effects of varying link capacity and PU detection errors on TCP throughput were studied in a dynamic spectrum access network with a centralized base station [32]. However, this study did not consider the ad hoc nature of the communications.

For CRAHNs, existing works mainly regard the adaptation of TCP to the dynamic radio environment by considering cognitive cycle operations. The transport layer has also been investigated in CRSNs.

In Ref. [35], it was pointed out that existing WSN transport layer solutions are not suitable for CRSNs, because they do not consider the additional delay caused by spectrum sensing and spectrum mobility. These were indicated as open research issues for the CRSN transport layer. It was concluded that to propose novel transport layer solutions, packet losses and delays due to the cognitive cycle must be considered together with the energy constraints of CRSN nodes.

3.4.2.3 WMSNs Multimedia communication in WSN requires the transportation of a great volume of data with high transmission rate and bounded delay. The high volume of data produced by multimedia applications results in buffer overflow and congestion. Queue-based Congestion Control Protocol with Priority Support (QCCP-PS) [36] was proposed for fair and reliable transport in WMSNs. In this protocol, queue length is used as a measure of congestion degree, and rate assignment to a traffic source is done according to its priority index and congestion degree. The priority index is used to ascertain the priority of the traffic sources in WMSNs. However, this approach does not consider spectrum-aware communication and hence it cannot be directly applied to CR ad hoc and sensor networks.

3.4.3 Open Research Issues for the Transport Layer

In preceding subsections, we investigated the challenges for multimedia communication in CR ad hoc and sensor networks. Furthermore, we surveyed the important existing transport layer solutions in WSNs, CRAHNs, CRSNs, and WMSNs.

Transport layer protocols for WSNs cannot be applied to CR and ad hoc networks due to lack of support for DSA functionalities. CRAHN and CRSN solutions do not satisfy the unique requirements of multimedia communication such as bounded delay, jitter control, and smooth variation of throughput. WMSN transport layer solutions satisfy multimedia communication requirements; however, they do not consider spectrum awareness. Hence, the open research issues can be itemized as follows:

- In light of the above discussions, it is necessary to develop novel transport layer algorithms to satisfy the requirements of multimedia communication and DSA functionalities.

The proposed transport protocol should be adaptive to licensed user activities and cognitive cycle operations to deliver multimedia data in an energy-efficient and timely manner.
- Varying network capacity is detrimental for multimedia communication. Spectrum handoff results in a change of operating channel parameters, and link capacity changes accordingly. Hence, licensed user behavior can be estimated to prevent frequent spectrum handoff and deterioration of multimedia communication.
- A novel congestion detection mechanism should be developed that considers the extra delay caused by cognitive cycle operations such as spectrum sensing and spectrum handoff.
- A novel rate control mechanism should be designed to decrease buffer overflow and retransmission. Rate control in CR ad hoc and sensor networks should consider the varying licensed user activity, spectrum-sensing duration, and spectrum handoff delay.
- In addition to reliable transport protocols for multimedia communication, they should support real-time communication to satisfy multimedia application deadlines.
- Due to energy constraints, the limited capabilities of sensor nodes, and the dynamic spectrum environment and DSA capability, cross-layer communication protocols should be investigated to support energy-efficient and reliable multimedia communication in CRSNs.

3.5 The Network Layer in Multimedia Communication for Cognitive Radio Ad Hoc and Sensor Networks

The network layer provides the connection between the transport layer and the MAC layer. It mainly provides routes from the source to destination that satisfy certain QoS requirements such as energy efficiency and delay. In wireless ad hoc networks, multihop communication is utilized to save energy and to maintain high-quality communication links. Hence, efficient network layer solutions are imperative for establishing energy-efficient routes and providing information to the transport layer.

3.5.1 Limitations and Challenges

Multimedia communication in CR sensor and ad hoc networks has some inherent limitations. Furthermore, spectrum awareness of wireless nodes poses extra challenges in the network layer.

Multimedia communication has stringent delay requirements and requires high bandwidth. Furthermore, there are limitations on multimedia communication in WSNs as stated in Ref. [3], which are summarized as follows. Multimedia communication has a query-driven or event-driven data delivery model. Continuous data delivery drains the energy of wireless nodes. Source aggregation and in-network processing are not efficient methods for multimedia communication due to the anisotropic nature of video signals. Furthermore, the cost due to the multimedia information gathering and processing in local communication should be limited. Multimedia communication should also satisfy a certain level of QoS. This limitation could be overcome by utilizing multipath communication.

The limitations of multimedia communication are amplified by the unique features of CR. For example, data delivery models must be revisited for the spectrum sensing and spectrum handoff mechanisms. In addition, the dynamic radio environment poses challenges for multipath communication, because multipath routes between an event and the sink may be disrupted by the activities of PUs. Predetermined routes may break due to dynamic channel conditions.

3.5.2 Existing Network Layer Solutions

3.5.2.1 WSNs Wireless sensor nodes have energy and processing constraints, hence energy efficiency is the main consideration of routing protocols in WSNs. The route is established and maintained at the network layer to maximize the lifetime of the WSN. Furthermore, an excessive number of sensor nodes is deployed in sensor networks; classical IP addressing would be an overhead for energy-constrained WSN regimes. Hence, attribute addressing is used such that the node having the attributed data communicates to the sink over multiple hops.

Routing protocols in WSNs are classified into three categories—flat, hierarchical, and location-based [37]. In flat network routing, all the nodes in the network have the same role and collaboratively send observation data to the sink. Hierarchical network routing employs clustering; cluster heads behave as source aggregation nodes and decrease message content to provide energy efficiency. In geographical network routing, each node in the network knows its position by triangulation and localization techniques. The source packets are routed toward the sink with the location information.

Sensor protocols for information via negotiation (SPIN) [38] is a flat network routing protocol where every node sends an advertisement (ADV) message to announce its data packet. The node needing this packet sends a request (REQ) message to the node, and the node with the data sends it with a DATA message. This protocol is not suitable for multimedia communication in CR ad hoc and sensor networks because it does not guarantee the delivery of data packets, which degrades the performance of the multimedia communication. Furthermore, Directed Diffusion [39] is a query-based data-centric routing protocol where the sink sends an interest signal to request information from the source nodes. Source nodes set up gradients toward the sink and the sink reinforces some paths according to QoS requirements. Source nodes send the information from reinforced paths. This process requires end-to-end control data transfer, hence the routing discovery process may increase delay bound. By contrast, the multipath support satisfies the multipath limitation of multimedia communication. A variant of Directed Diffusion, Rumor Routing [40], which decreases energy consumption by avoiding flooding the queries, does not provide multipath support; therefore, it is not suitable for multimedia communication.

Hierarchical routing for multimedia communication has some drawbacks because of the communication and processing burden of multimedia data at the cluster head and poor channel quality between cluster heads, which is the result of a high communication range between the cluster heads. In Low-Energy Adaptive Clustering Hierarchy (LEACH) [41], the data collected by the cluster members are gathered at the cluster head, and an aggregation technique is applied. LEACH proposes a routing architecture such that the aggregated data at each cluster head are conveyed collaboratively by all cluster heads. It offers scalability, lifetime improvement, and energy efficiency. Due to the aggregation limitations of multimedia data and the poor quality of the communication link between cluster heads, this architecture is not suitable for multimedia communication.

On-the-fly event-driven clustering was proposed in Ref. [42], where it is called "Event-to-Sink Directed Clustering" (ESDC). Upon detection of an event, a routing corridor is established between the event and the sink. The data packets are routed by the cluster heads. Multimedia packets are large packets and their processing, that is, aggregation, requires a large memory size, which is not suited to the limited capabilities of sensor nodes. Hence, this approach has a disadvantage for multimedia communication.

Clustering-based routing is very promising for WSNs; however, if the network is homogenous, the multimedia applications cannot use the benefits of hierarchical routing. By contrast, if the cluster heads are more powerful nodes that are capable of processing multimedia data, hierarchical routing becomes an efficient means for providing network layer solutions in multimedia applications.

In another approach, the event packets are propagated directly towards the sink, if the direction of the base station is known. First, the node broadcasts its packet to its one-hop neighbors; the node with the least cost to the base station continues the data flow. In this approach, known as the "Minimum Cost Forwarding Algorithm" (MCFA) [43], each node knows the cost to reach the base station through messages sent by the base station. This protocol is convenient for multimedia communication, since the multimedia data are forwarded to the base stations upon the detection of an event without any delay for setting up routes.

In location-based networks, routing is performed according to the location information. For example, in Geographical and Energy Aware Routing (GEAR) [44], location-aware and energy-aware nodes utilize a heuristic approach for selecting a neighbor to route the packets toward the sink. In Ref. [45], energy minimization is achieved through location information and relay region determination accordingly. Minimum energy paths are established beforehand; hence, the topology changes and reconfiguration of WSNs may lead to inefficiency of the routing protocol. However, location-based networks set up routes with smaller delay and reduced energy consumption. Hence, it is more appropriate for multimedia communication.

Routing protocols can also be divided into five categories, according to protocol operation: negotiation-based, multipath, query-based, QoS-based, and coherence-based protocols [37]. In negotiation-based protocols, negotiation is performed between nodes to prevent duplication and redundancy before data communication. In multipath communication, the same data are sent along different paths to increase reliability and bandwidth. Query-based protocols start data transmission and set up routes upon queries sent by the sink. QoS-based routing protocols establish routes to satisfy certain parameters such as delay, throughput, jitter, and so on. In coherence-based routing protocols, the raw data are processed at the nodes. According to this characterization, multipath and QoS routing protocols are suitable for multimedia communication because it requires reliability, high throughput, and low delay bound and jitter.

QoS-based protocols set up and maintain the routes according to certain metrics such as link delay, bandwidth, energy efficiency, residual energy, or a weighted combination of some of these. In Ref. [46], an energy-aware QoS routing protocol is proposed. It considers the end-to-end delay to establish QoS routes for real-time data delivery. According to a cost function—which is evaluated by transmission energy, error rate, and residual energy—the least-cost path is selected with delay constraint. Real-Time Power-Aware Routing (RPAR) [47] addresses the challenge of energy-efficient communication with a certain QoS requirement—specifically, delay. RPAR is achieved by adapting the transmission power of the nodes and routing decisions dynamically, based on packet deadlines. It supports real-time communication in WSNs. This type of protocol is very promising for multimedia communication in CR ad hoc and sensor networks because it satisfies a certain QoS requirement—delay, with energy-efficiency constraints. However, such protocols do not support CR functionality. Hence, they need to be modified to adapt to spectral changes.

Table 3.3 outlines routing protocols in WSNs and their advantages and disadvantages for realizing multimedia communication in CR ad hoc and sensor networks.

3.5.2.2 Cognitive Radio Ad Hoc and Sensor Networks The dynamic spectral environment and cognitive cycle operations of wireless nodes make routing a very challenging task in CRNs. The flexibility

Table 3.3 Routing Protocols in Wireless Sensor Networks

PROTOCOL	MULTIMEDIA COMMUNICATION ISSUES
SPIN [38]	– Flat network routing – No guarantee of data delivery
Directed Diffusion [39]	+ Query-based data-centric routing + Multipath support – Delay due to end-to-end control data
Rumor Routing [40]	+ A variant of directed diffusion [39] + No flooding of queries – No multipath support
LEACH [41]	+ Scalability and lifetime improvement – Poor link quality between cluster heads
ESDC [42]	+ On-the-fly event-driven clustering + Routing corridor between event and sink
GEAR [44]	+ Route selection according to energy and location
[45]	+ Relay region determination through location information
MCFA [43]	+ Routing toward the sink + No delay to set up routes to the sink
[46]	+ Energy-aware QoS-based routing + QoS routes for real-time data delivery
RPAR [47]	+ Energy-efficient routing with low delay + Routing decisions according to packet deadline

Note: Advantages and disadvantages are indicated by + and –, respectively.

in spectrum access brings about complexity in the protocol design. The main challenges of the routing multihop CRN are spectrum awareness, setup of quality routes, and route maintenance/repair [48]. Furthermore, routing protocols in CRNs depend on either full-spectrum knowledge or local-spectrum knowledge.

With full-spectrum knowledge in CRNs, optimal routes can be set up. In practical networks, full knowledge may not be achievable; however, optimal routes can be used as a benchmark for the routing protocols. In the routing process, avoidance of interference with PUs is a requirement; hence, routing protocols establish routes to not interfere with licensed users. In Ref. [49], spectrum information is known *a priori* and mixed integer linear programming is used to provide fair routing for different traffic demands by considering spectrum sharing and flow routing with interference considerations.

Spectrum-Aware Opportunistic Routing [50] was proposed to employ multipath transmissions and QoS guaranteed throughput with decreased delay. SAOR creates a spectrum map from local sensing observations and opportunistic routing. Furthermore, opportunistic cognitive routing was proposed in Ref. [51]. CRs select the relay node based on the location information and channel usage statistics. A new metric, cognitive transport throughput, is defined to assess the potential relay gain.

Link scheduling is important for CRNs to decrease interference. The routing problem in multihop CRNs has been studied with link scheduling [52]. Scheduling and routing are performed alongside the unpredictable activities of PUs. They try to perform opportunistic spectrum access, scheduling, and multipath multihop routing in CRNs by minimizing network resource usage. The spectrum vacancy of a licensed band is modeled as a random variable to estimate the channel availability. Ref. [52] is an important study for realizing multimedia communication in CR ad hoc and sensor networks.

Spectrum-Aware On-Demand Routing protocol is a reactive protocol that balances channel switching and spectrum sharing through routing and frequency band selection [53]. Routes are selected according to their effectiveness in terms of spectrum switching delay and back-off delay. Ma et al. [54] also propose an on-demand routing and channel assignment protocol that does not utilize a common control channel (CCC). The protocol also decreases back-off delay and the overhead due to channel switching and tries to avoid the deafness problem due to channel switching. Reactive protocols are more advantageous than proactive protocols because the dynamic nature of the environment renders obsolete the routes established beforehand. Hence, on-demand (reactive) routing protocols are more applicable to CR ad hoc and sensor networks.

Enabling multimedia communication of WMSNs over CRSNs poses challenges due to the distributed nature of CRSNs and the dynamics of network and spectrum usage. Hence, routing protocols should provide rate adaptation with opportunistic bandwidth, as well as delay and jitter control [55]. Spectrum-Aware Clustering Protocol for Energy Efficient Routing (SCEEM) [55] was proposed such that clustering manages dynamic spectrum access and QoS routing for multimedia communication in CRSNs. The contributions of

Table 3.4 Routing Protocols in Cognitive Radio Networks

PROTOCOL	MULTIMEDIA COMMUNICATION ISSUES
Joint Spectrum and Fair Routing Protocol [49]	+ Can be considered a benchmark
	+ Fair routing for different traffic demands
	– Spectrum information known *a priori*
SAOR [50]	+ Multipath opportunistic routing
	+ QoS guaranteed throughput with decreased delay
	+ Local spectrum sensing
Opportunistic Cognitive Routing (OCR) [51]	+ Relay node selection based on location information and channel usage statistics
Joint Routing and Link Scheduling Protocol [52]	+ Joint routing and scheduling
Spectrum-aware On-Demand Routing Protocol [53]	+ Balances spectrum sharing and channel switching
	+ Considers spectrum switching and back-off delay
Multi-hop single transceiver CRN Routing Protocol [54]	+ Route selection according to energy and location
SCEEM [55]	+ Smooth delivery of multimedia data
	+ Clustered QoS routing

Note: Advantages and disadvantages are indicated by + and –, respectively.

SCEEM are smooth multimedia data delivery by the isolation of time and frequency variability of the spectrum and energy efficiency with QoS-aware routing.

Table 3.4 outlines several routing protocols in CRNs and CRSNs and their advantages for multimedia communication in CR ad hoc and sensor networks.

CRN routing protocols consider delay, throughput, and licensed user activities; however, they do not deal with the challenge of energy efficiency. Delay and throughput considerations are compatible with the requirements of multimedia communication. By contrast, these protocols need appropriate modifications such as energy efficiency to be utilized in CR ad hoc and sensor networks.

3.5.2.3 WMSNs New routing solutions have been proposed to satisfy the unique requirements of multimedia communication in WMSNs. Readers may refer to the comprehensive survey about routing protocols in WMSN presented by Abazeed et al. [56]. In this section, we present some of the recent proposals for routing in WMSNs.

Li et al. [57] extend directed diffusion to reinforce multiple paths with a high quality of links and low latency. This protocol offers better throughput with lower delay than directed diffusion. Furthermore, Ref. [58] proposes a multichannel, multipath routing protocol in which routes are determined by adjusting the required bandwidth dynamically and by the differentiation of delay that is proportional to the path length for real-time data. Ref. [59] provides the necessary bandwidth for multimedia applications by establishing disjoint routes and avoiding intersession and intrasession interference. One path is built for each session; if there is congestion or lack of bandwidth, additional paths are established. Interfering nodes are forced to be passive to decrease interference and unnecessary energy consumption.

A routing protocol has also been proposed for multimedia streaming, which requires delay bound and high bandwidth demand [60]. This protocol is an online multipath routing protocol for WMSNs. Route decisions are made at each hop using location information. Two schemes are used to achieve energy efficiency and load balancing: a smart greedy forwarding scheme that selects the most appropriate node and a walk-back scheme to bypass network holes.

Table 3.5 outlines the state-of-the-art routing protocols and their advantages for realizing multimedia communication in CR ad hoc and sensor networks. In general, the proposed routing solutions in the literature employ location-based and multipath routing. They support traffic demanding high bandwidth and low delay. However, the proposed solutions do not consider the dynamic radio environment, which is a missing issue for multimedia communication in CR ad hoc and sensor networks.

Table 3.5 Routing Protocols in Wireless Multimedia Sensor Networks

PROTOCOL	MULTIMEDIA COMMUNICATION ISSUES
Delay Constrained High Throughput Protocol [57]	+ Reinforces multiple paths with high quality links with low latency
QoS Routing Protocol [58]	+ Multichannel multipath routing + Real-time data delivery
Maximally radio-disjoint multipath routing Protocol [59]	+ Disjoint route establishment + Additional paths in case of congestion and lack of bandwidth

Note: Advantages and disadvantages are indicated by + and –, respectively.

3.5.3 Open Research Issues for the Network Layer

Numerous efforts focus on network layer solutions in WSNs, CRNs, and WMSNs. These efforts mainly concentrate on routing protocols. Multimedia communication requirements are satisfied by WMSN routing solutions. However, they must also overcome the challenges of CR and sensor networks. Open research issues in the network layer can be itemized as follows.

- Back-off delay and channel-switching delay must be taken into account to provide delay-bounded path establishment. Additional delay occurs due to licensed user activities in CR ad hoc and sensor networks.
- The dynamic radio environment causes changes in spectrum opportunities; hence, reactive protocols would be more appropriate for multimedia communication in CR ad hoc and sensor networks. This type of protocol should be designed.
- Due to the dynamic spectral environment, established paths can become obsolete. Hence, multipath protocols are appropriate for multimedia communication, and there is no existing study of multipath protocols in CR ad hoc and sensor networks.

3.6 The MAC Layer in Multimedia Communication for Cognitive Radio Ad Hoc and Sensor Networks

The MAC layer is mainly responsible for the channel access mechanism, scheduling, buffer management, and error control mechanisms in wireless networks. The channel access mechanism affects the delay and reliability of the packet; scheduling is used to schedule packet transmission in the medium to decrease collision probability. The bursty nature of multimedia communication requires efficient buffer management policies to decrease buffer overflows. If the received packet has errors, the transmitter simply resends the packet in ARQ scheme. However, in FEC scheme, the receiver can detect and recover the errors in the packet without any retransmission. Due to the real-time data delivery requirement of multimedia communication, the ARQ scheme may not be used, because it requires retransmission of the packet.

Energy-aware MAC layer solutions are necessary to satisfy the unique requirements of the sensor network paradigm as well as the requirements of multimedia communication. Novel techniques are required for adaptation to the dynamic spectral environment in CR ad hoc and sensor networks. Access to the medium, error control, and packet transmission reliability are highly influenced by the requirements of multimedia communication, sensor networks, and CR. In this section, we survey MAC layer solutions in the literature and how they can be extended to multimedia communication in CR ad hoc and sensor networks.

3.6.1 Challenges and Requirements

The major requirements of MAC protocols in sensor networks are real-time or QoS requirements, decentralized operation, power awareness, and flexibility to different kinds of applications [61]. The main requirement of multimedia communication in the MAC layer is packet transmission with minimal error, delay, and collision. Contention in the medium access may result in collision and, hence, delay and energy consumption. Latency, delay, and jitter are the vital QoS parameters for multimedia communication.

The most important challenge for multimedia communication in CR ad hoc and sensor networks is intermittent communication due to cognitive cycle functions. These fundamental operations result in delay and channel quality degradation and, hence, a decrease in communication quality for multimedia communication. In CR communication, there is a nonzero probability of colliding with the packets of PUs, as well as with the packets of SUs. Furthermore, there are many more parameters that need to be taken into account to realize multimedia communication in CR ad hoc and sensor networks.

3.6.2 Existing MAC Layer Solutions

There have been some surveys on the MAC layer in WSNs [62] and CRNs [63,64]; however, they do not explain how to enable efficient multimedia communication in dynamic radio environments. In this section, in addition to explaining the state-of-the-art solutions for the MAC layer, we emphasize the changes necessary to realize multimedia communication in CR ad hoc and sensor networks.

3.6.2.1 WSNs Many MAC layer protocols have been proposed with different objectives for WSNs. However, they do not satisfy the requirements of multimedia communication and the challenges posed by CR and sensor networks. In sensor networks, the primary concerns are energy efficiency and scalability; however, throughput and latency attributes are secondary concerns [62]. They must all be treated as primary concerns for multimedia communication.

MAC protocols can be divided into two categories: contention-free protocols and contention protocols. The contention-free protocols are time division multiple access (TDMA), frequency division multiple access (FDMA), and code division multiple access (CDMA). By contrast, contention protocols do not reserve resources; they are allocated to network users on demand. The carrier sense multiple access (CSMA) protocol and its variants are used for medium access control in sensor networks as contention protocols.

The main attributes of MAC protocols are energy efficiency, scalability and adaptivity, channel utilization, latency, throughput, and fairness [65]. In contention-free protocols, scheduling is organized by time, frequency, or code domain to eliminate interference and collisions. TDMA protocols [66] support low duty cycle operation, which decrease idle listening; however, TDMA protocols have scalability issues and do not support distributed functionality. Furthermore, TDMA protocols require global synchronization, which is a burden for sensor networks. The clustering approach has been proposed to resolve the scalability issue and scheduling. LEACH [41] proposes TDMA scheduling for its cluster and cluster members. Furthermore, the protocol in Ref. [67] uses different subchannels by frequency division and code division and, hence, utilize FDMA and CDMA. Reservation of network resources decreases collision, delay, and jitter; increases throughput; and provides real-time guarantees, which are fundamental requirements of multimedia communication [2]. However, the centralized control and the complexity of the protocols are the disadvantages.

The contention-based protocols use carrier-sensing and collision avoidance mechanisms [61]. The challenge of contention-based protocols is collision due to insufficient coordination. For example, two disparate nodes may send packets to the same receiver and there may be collision at the receiver node. Such an event is termed a *hidden node terminal problem*. Multiple access with collision avoidance (MACA) [68] eliminates this problem with extra control signals, request to send (RTS) and clear to send (CTS). It uses three-way handshaking to coordinate communication. The control signals may collide; however, the collision probability is low due to the small control packet size. Variants of this approach exist, such as multiple access with collision avoidance for wireless (MACAW) [69] and MACA with piggybacked reservation (MACA/PR) [70].

Institute of Electrical and Electronics Engineers (IEEE) 802.11 [71] is a standardized protocol for wireless LAN. It combines the features of CSMA/CA, MACA, and MACAW. The IEEE 802.11 distributed coordination function is designed for ad hoc networks. It utilizes the CSMA/CA and RTS/CTS handshaking mechanism. If the medium is idle for a predefined interval, the node is allowed to transmit. If the channel is sensed to be busy, a random back-off interval is been chosen uniformly from the contention window interval. Furthermore, ACK and virtual carrier sensing by network allocation vector are used. The main disadvantage of this protocol is the listening to the channel during the back-off period, which is energy-consuming.

Hybrid schemes combine the contention-based period with the contention-free period. Nodes in a neighborhood first contend for reservation of the medium. After the reservation period, the reserved communication resources are used by the nodes. CSMA approaches do not require clock synchronization or global knowledge about the network. However, contention-free protocols are not tolerant to sensor node failures, and clock synchronization and a central controller are necessary. Dynamic topology changes in sensor networks may be costly due to the global rearrangement of schedules. Low contention in sensor networks causes lower channel utilization and higher delay than CSMA, since the node must wait for its TDMA schedule to communicate; by contrast, in CSMA the node communicates when the channel is sensed to be idle [72]. Hence, Rhee et al. proposes Z-MAC, a hybrid MAC protocol that combines the strengths of CSMA and TDMA while decreasing their inefficiencies. The main advantage of Z-MAC is the adaptability of network contention level. With low contention it becomes TDMA; with high contention it becomes TDMA.

From the perspective of multimedia communications, collision-free protocols can provide real-time guarantees, as well as tolerable delay

Table 3.6 MAC Protocols in Wireless Sensor Networks

PROTOCOL	TYPE	TECHNIQUE	DISADVANTAGE
Low Power TDMA [66]	Contention-free	TDMA	Global synchronization
LEACH [41]	Contention-free	TDMA	Clustered structure
Self Organization Protocol [67]	Contention-free	CDMA-FDMA	Lack of scalability
Real time communication and coordination protocol [61]	Contention	Carrier sensing	Coordination
MACA [68]	Contention	Carrier sensing	Collision
IEEE 802.11 [71]	Contention	CSMA/CA	Idle listening
Z-MAC [72]	Hybrid	CSMA-TDMA	—

bounds and desirable throughput by organizing the communication schedule. On the other hand, this type of protocol is not scalable, and they do not adapt to the dynamic environment of sensor networks. Contention-based protocols cannot provide real-time guarantees due to collisions. Hybrid protocols can be used for multimedia communication, since the disadvantages of contention-based and contention-free protocols are minimized by combining them. Hybrid mechanisms are more resilient to scalability, energy consumption, and collisions.

Table 3.6 outlines MAC protocols, their types, techniques used, and disadvantages for multimedia communication in CR ad hoc and sensor networks.

3.6.2.2 Cognitive Radio Networks One of the most fundamental challenges for designing MAC protocols in CRNs is spectrum management. Medium access is highly affected by spectrum-sensing results, and communication between nodes is interrupted by spectrum handoff and spectrum sensing. Resource allocation is performed to eliminate collision between CRs and to avoid harmful interference with PUs.

MAC protocols can be split into two categories: Direct Access Based (DAB) and DSA [64]. In DAB protocols, every node optimizes its own goals and resource allocation is performed by sender–receiver handshake. DSA protocols try to achieve global network optimization.

Coordination among the nodes is a difficult challenge in CRNs. Two nodes within the communication region of each other communicate if they have a common channel. Agreement on communication channel can be reached on an in-band or out-of-band channel. In-band channel negotiation utilizes data channels; by contrast, out-of-band channel negotiation utilizes a different channel for negotiation and sharing of the spectrum-sensing results. Generally, the out-of-band approach is used in MAC protocols. The out-of-band approach can be divided into CCC and split phase (SP). CCC uses a common control channel for signaling [73], and SP allows the transceiver on each node to switch between control phase and data phase in a time frame [74]. CCC needs two transceivers and handles synchronization problems; SP is cost-efficient but reduces network utilization due to the separate control phase. The CCC approach is much more convenient than the SP approach, because it overcomes the problem of synchronization and system inefficiency. However, it is not practical to place two transceivers on one sensor node. Multimedia QoS requirements can be satisfied in a more robust way by the CCC approach. Nevertheless, it should be noted that it is not feasible to guarantee CCC among the network members due to the dynamic nature of the spectral environment in CRNs. By contrast, the Frequency Hopping Sequence (FHS) approach utilizes in-band control channel, and sensing interruptions do not affect system performance by hopping the frequency to enable communication [75]. FHS generally satisfies the throughput and QoS requirements of multimedia communication.

Contention in CRNs can be avoided by centralized or distributed algorithms. As classified in Ref. [63], there are three kinds of CR MAC protocols: random access, time-slotted, and hybrid. In random access MAC protocols, CRs utilize CSMA/CA as architecture for accessing the medium for control and data packets. A centralized carrier-sensing CSMA-based mechanism has been proposed for CRNs coexisting with a primary system. It utilizes in-band signaling, and CR nodes have single transceivers. In time-slotted MAC protocols, time is synchronized and slots for data and control channels are reserved. In cognitive MAC (C-MAC) [76], time is divided into superframes which contain slots for beacon period and data transmission period. It uses multiple transceivers to enable high throughput and resilience to the dynamic spectrum environment. A rendezvous channel is used for node coordination. A back-up channel is determined from out of band in case a PU appears on the rendezvous channel. In hybrid protocols, medium access is achieved by utilizing random access and time-slotted schemes. For example, in opportunistic spectrum MAC (OS-MAC) [77], there are predetermined intervals for communication coordination. CCC is used to exchange control packets, and each node is a member of a cluster.

From a multimedia communication perspective, MAC protocols face challenging difficulties. Existing MAC layer approaches do not support real-time, seamless communication as a result of PU and high-throughput requirements, jitter, and delay bound. An important MAC protocol has been proposed to support multimedia communication in CRNs [78]. It is a distributed QoS-aware MAC protocol for multichannel CRNs. Based on channel usage statistics, the channels to be sensed and the data channels are determined to satisfy certain QoS requirements. For various traffic types, spectrum-sensing duration is changed to enhance QoS provisioning. Cai et al. [78] offer a priority-based spectrum access scheme for heterogeneous traffic to further improve QoS. MAC protocols for multimedia applications must consider QoS requirements such as traffic, PU activities, real-time communication, and delay bound. Furthermore, cognitive cycle operations must be considered for medium access such that spectrum-sensing results and spectrum decision are fundamental operations that affect the performance of the communication.

3.6.3 Open Research Issues for the MAC Layer

There are many approaches for the MAC layer in WSNs that satisfy the requirements of multimedia communication. In contrast, the protocols proposed for CRNs are mainly concerned with decreasing interference with PUs and increasing channel utilization. These protocols generally do not consider multimedia communication. Hence, there is a need to design a MAC protocol to realize multimedia communication in CR multimedia ad hoc and sensor networks.

The proposed MAC protocol must minimize the delay caused by spectrum handoff, as well as contention. Furthermore, it must offer a high throughput against changing the operating channel and its condition. Hence, designing a MAC protocol to satisfy the requirements of multimedia communication is a challenging task.

3.7 Conclusions

In this chapter, we surveyed the existing literature on multimedia communication in WSNs, CRNs, CRSNs, and WMSNs. We also investigated the protocols proposed for wireless networks and pointed out their advantages and disadvantages for realizing multimedia communication in CR ad hoc and sensor networks. Existing approaches were scrutinized from the perspective of network layers. The challenges posed by multimedia communication and the CR ad hoc and sensor network paradigm were clearly indicated for each network layer. Open research avenues for realizing multimedia communication over CR ad hoc and sensor networks were presented.

Acknowledgment

This work was supported by The Scientific and Technological Research Council of Turkey (TUBITAK) under grant 110E249.

References

1. Akyildiz, I.F., Su, W., Sankarasubramaniam, Y., Cayirci, E. Wireless sensor networks: A survey. *Computer Networks* 38(4), 393–422, 2002.
2. Misra, S., Reisslein, M., Xue, G. A survey of multimedia streaming in wireless sensor networks. *Communications Surveys & Tutorials, IEEE* 10(4), 18–39, 2008.
3. Gürses, E., Akan, O.B. Multimedia communication in wireless sensor networks. In: *Annales des Télécommunications*, volume 60, pp. 872–900. Springer, 2005.
4. Akyildiz, I.F., Lee, W.Y., Vuran, M.C., Mohanty, S. Next generation/dynamic spectrum access/cognitive radio wireless networks: A survey. *Computer Networks* 50(13), 2127–2159, 2006.
5. Akan, O.B., Karli, O., Ergul, O. Cognitive radio sensor networks. *Network, IEEE* 23(4), 34–40, 2009.
6. Haykin, S. Cognitive radio: Brain-empowered wireless communications. *IEEE Journal on Selected Areas in Communications* 23(2), 201–220, 2005.
7. Akyildiz, I.F., Melodia, T., Chowdhury, K.R. A survey on wireless multimedia sensor networks. *Computer Networks* 51(4), 921–960, 2007.
8. Vuran, M.C., Akan, O.B., Akyildiz, I.F. Spatio-temporal correlation: Theory and applications for wireless sensor networks. *Computer Networks* 45(3), 245–259, 2004.
9. Wiegand, T., Sullivan, G.J. The H.264/AVC video coding standard. *IEEE Signal Processing Magazine* 24(2), 148–153, 2007.
10. Ohm, J., Sullivan, G.J. High efficiency video coding: The next frontier in video compression [standards in a nutshell]. *Signal Processing Magazine, IEEE* 30(1), 152–158, 2013.
11. Slepian, D., Wolf, J.K. Noiseless coding of correlated information sources. *IEEE Transactions on Information Theory* 19(4), 471–480, 1973.

12. Girod, B., Aaron, A.M., Rane, S., Rebollo-Monedero, D. Distributed video coding. *Proceedings of the IEEE* 93(1), 71–83, 2005.
13. Aaron, A., Rane, S., Zhang, R., Girod, B. Wyner–Ziv coding for video: Applications to compression and error resilience. In: *Proceedings of Data Compression Conference*, pp. 93–102. IEEE, Snowbird, UT, 2003.
14. Aaron, A., Setton, E., Girod, B. Towards practical Wyner–Ziv coding of video. In: *Proceedings of International Conference on Image Processing, ICIP*, volume 3, pp. III–869. IEEE, Barcelona, Spain, 2003.
15. Brites, C., Pereira, F. Correlation noise modeling for efficient pixel and transform domain Wyner–Ziv video coding. *IEEE Transactions on Circuits and Systems for Video Technology* 18(9), 1177–1190, 2008.
16. Brites, C., Pereira, F. An efficient encoder rate control solution for transform domain Wyner–Ziv video coding. *IEEE Transactions on Circuits and Systems for Video Technology* 21(9), 1278–1292, 2011.
17. Yu, F.R., Sun, B., Krishnamurthy, V., Ali, S. Application layer QoS optimization for multimedia transmission over cognitive radio networks. *Wireless Networks* 17(2), 371–383, 2011.
18. Dastpak, A., Liu, J., Hefeeda, M. Video streaming over cognitive radio networks. In: *Proceedings of the 4th Workshop on Mobile Video*, pp. 31–36. ACM, Chapel Hill, NC, 2012.
19. Zhang, H., Zhang, Z., Chen, X., Yin, R. Energy efficient joint source and channel sensing in cognitive radio sensor networks. In: *IEEE International Conference on Communications (ICC)*, pp. 1–6. IEEE, Kyoto, Japan, 2011.
20. Wang, C., Sohraby, K., Li, B., Daneshmand, M., Hu, Y. A survey of transport protocols for wireless sensor networks. *IEEE Network* 20(3), 34–40, 2006.
21. Lee, W.Y., Akyildiz, I.F. Optimal spectrum sensing framework for cognitive radio networks. *IEEE Transactions on Wireless Communication* 7(10), 3845–3857, 2008.
22. Hull, B., Jamieson, K., Balakrishnan, H. Mitigating congestion in wireless sensor networks. In: *Proceedings of the 2nd International Conference on Embedded Networked Sensor Systems*, pp. 134–147. ACM, Baltimore, MD, 2004.
23. Wan, C.Y., Eisenman, S.B., Campbell, A.T. CODA: Congestion detection and avoidance in sensor networks. In: *Proceedings of the 1st International Conference on Embedded Networked Sensor Systems*, pp. 266–279. ACM, Los Angeles, CA, 2003.
24. Ee, C.T., Bajcsy, R. Congestion control and fairness for many-to-one routing in sensor networks. In: *Proceedings of the 2nd International Conference on Embedded Networked Sensor Systems*, pp. 148–161. ACM, Baltimore, MD, 2004.
25. Stann, F., Heidemann, J. RMST: Reliable data transport in sensor networks. In: *Proceedings of the 1st IEEE International Workshop on Sensor Network Protocols and Applications*, pp. 102–112. IEEE, Anchorage, Alaska, 2003.

26. Felemban, E., Lee, C.G., Ekici, E. MMSPEED: Multipath multi-SPEED protocol for QoS guarantee of reliability and. timeliness in wireless sensor networks. *IEEE Transactions on Mobile Computing* 5(6), 738–754, 2006.
27. Wan, C.Y., Campbell, A.T., Krishnamurthy, L. PSFQ: A reliable transport protocol for wireless sensor networks. In: *Proceedings of the 1st ACM International Workshop on Wireless Sensor Networks and Applications*, pp. 1–11. ACM, Atlanta, GA, 2002.
28. Sankarasubramaniam, Y., Akan, O.B., Akyildiz, I.F. ESRT: Event-to-sink reliable transport in wireless sensor networks. In: *Proceedings of the 4th ACM International Symposium on Mobile Ad Hoc Networking & Computing*, pp. 177–188. ACM, Annapolis, Maryland, 2003.
29. Paek, J., Govindan, R. RCRT: Rate-controlled reliable transport for wireless sensor networks. In: *Proceedings of the 5th International Conference on Embedded Networked Sensor Systems*, pp. 305–319. ACM, Sydney, Australia, 2007.
30. Gungor, V.C., Akan, O.B., Akyildiz, I.F. A real-time and reliable transport $(RT)^2$ protocol for wireless sensor and actor networks. *IEEE/ACM Transactions on Networking* 16(2), 359–370, 2008.
31. Chowdhury, K.R., Di Felice, M., Akyildiz, I.F. TCP-CRAHN: A transport control protocol for cognitive radio ad hoc networks. *IEEE Transactions on Mobile Computing* 12(4), 790–803, 2013.
32. Slingerland, A.M.R., Pawelczak, P., Venkatesha Prasad, R., Lo, A., Hekmat, R. Performance of transport control protocol over dynamic spectrum access links. In: *2nd IEEE International Symposium on New Frontiers in Dynamic Spectrum Access Networks, DySPAN 2007*, pp. 486–495. IEEE, Dublin, Ireland, 2007.
33. Chowdhury, K.R., Di Felice, M., Akyildiz, I.F. TP-CRAHN: A transport protocol for cognitive radio ad-hoc networks. In: *IEEE INFOCOM 2009*, pp. 2482–2490. IEEE, Rio de Janerio, Brazil, 2009.
34. Luo, C., Yu, F.R., Ji, H., Leung, V.C.M. Cross-layer design for TCP performance improvement in cognitive radio networks. *IEEE Transactions on Vehicular Technology* 59(5), 2485–2495, 2010.
35. Bicen, A.O., Akan, O.B. Reliability and congestion control in cognitive radio sensor networks. *Ad Hoc Networks* 9(7), 1154–1164, 2011.
36. Yaghmaee, M.H., Adjeroh, D. A new priority based congestion control protocol for wireless multimedia sensor networks. In: *IEEE WoWMoM 2008*, pp. 1–8. IEEE, Newport Beach, CA, 2008.
37. Al-Karaki, J.N., Kamal, A.E. Routing techniques in wireless sensor networks: A survey. *IEEE Wireless Communications* 11(6), 6–28, 2004.
38. Kulik, J., Heinzelman, W., Balakrishnan, H. Negotiation-based protocols for disseminating information in wireless sensor networks. *Wireless Networks* 8(2/3), 169–185, 2002.
39. Intanagonwiwat, C., Govindan, R., Estrin, D., Heidemann, J., Silva, F. Directed diffusion for wireless sensor networking. *IEEE/ACM Transactions on Networking* 11(1), 2–16, 2003.

40. Braginsky, D., Estrin, D. Rumor routing algorithm for sensor networks. In: *Proceedings of the 1st ACM International Workshop on Wireless Sensor Networks and Applications*, pp. 22–31. ACM, Atlanta, GA, 2002.
41. Heinzelman, W.R., Chandrakasan, A., Balakrishnan, H. Energy-efficient communication protocol for wireless microsensor networks. In: *Proceedings of the 33rd Annual Hawaii International Conference on System Sciences*, pp. 1–10. IEEE, Hawaii, USA, 2000.
42. Bereketli, A., Akan, O.B. Event-to-sink directed clustering in wireless sensor networks. In: *Wireless Communications and Networking Conference, 2009, WCNC 2009.* IEEE, pp. 1–6. IEEE, Budapest, Hungary, 2009.
43. Ye, F., Chen, A., Lu, S., Zhang, L. A scalable solution to minimum cost forwarding in large sensor networks. In: *Proceedings of 10th International Conference on Computer Communications and Networks*, pp. 304–309. IEEE, Scottsdale, AZ, 2001.
44. Yu, Y., Govindan, R., Estrin, D. Geographical and energy aware routing: A recursive data dissemination protocol for wireless sensor networks. Technical report, *Technical Report UCLA/CSD-TR-01-0023*, UCLA Computer Science Department, Los Angeles, CA, 2001.
45. Rodoplu, V., Meng, T.H. Minimum energy mobile wireless networks. *IEEE Journal on Selected Areas in Communications* 17(8), 1333–1344, 1999.
46. Akkaya, K., Younis, M. An energy-aware QoS routing protocol for wireless sensor networks. In: *Proceedings of 23rd International Conference on Distributed Computing Systems Workshops*, pp. 710–715. IEEE, Providence, RI, 2003.
47. Chipara, O., He, Z., Xing, G., Chen, Q., Wang, X., Lu, C., Stankovic, J. et al. Real-time power-aware routing in sensor networks. In: *14th IEEE International Workshop on Quality of Service*, pp. 83–92. IEEE, New Haven, CT, 2006.
48. Cesana, M., Cuomo, F., Ekici, E. Routing in cognitive radio networks: Challenges and solutions. *Ad Hoc Networks* 9(3), 228–248, 2011.
49. Ma, M., Tsang, D.H.K. Joint spectrum sharing and fair routing in cognitive radio networks. In: *5th IEEE Consumer Communications and Networking Conference, CCNC 2008*, pp. 978–982. IEEE, Las Vegas, Nevada, 2008.
50. Lin, S.C., Chen, K.C. Spectrum aware opportunistic routing in cognitive radio networks. In: *IEEE Global Telecommunications Conference (GLOBECOM 2010)*, pp. 1–6. IEEE, Miami, FL, 2010.
51. Liu, Y., Cai, L.X., Shen, X. Spectrum-aware opportunistic routing in multi-hop cognitive radio networks. *IEEE Journal on Selected Areas in Communications* 30(10), 1958–1968, 2012.
52. Pan, M., Zhang, C., Li, P., Fang, Y. Joint routing and link scheduling for cognitive radio networks under uncertain spectrum supply. In: *INFOCOM, 2011 Proceedings IEEE*, pp. 2237–2245. IEEE, Shanghai, China, 2011.
53. Cheng, G., Liu, W., Li, Y., Cheng, W. Spectrum aware on-demand routing in cognitive radio networks. In: *2nd IEEE International Symposium on New Frontiers in Dynamic Spectrum Access Networks*, pp. 571–574. IEEE, Dublin, Ireland, 2007.

54. Ma, H., Zheng, L., Ma, X., Luo, Y. Spectrum aware routing for multi-hop cognitive radio networks with a single transceiver. In: *3rd International Conference on Cognitive Radio Oriented Wireless Networks and Communications, 2008. CrownCom 2008.*, pp. 1–6. IEEE, Singapore, 2008.
55. Shah, G.A., Alagoz, F., Fadel, E., Akan, O.B. A spectrum-aware clustering for efficient multimedia routing in cognitive radio sensor networks. *IEEE Transactions On Vehicular Technology* 63(7), 3369–3380, 2013.
56. Abazeed, M., Faisal, N., Zubair, S., Ali, A. Routing protocols for wireless multimedia sensor network: A survey. *Journal of Sensors* 1–11, 2013.
57. Li, S., Neelisetti, R., Liu, C., Lim, A. Delay-constrained high throughput protocol for multi-path transmission over wireless multimedia sensor networks. In: *International Symposium on a World of Wireless, Mobile and Multimedia Networks, WoWMoM 2008*, pp. 1–8. IEEE, Newport Beach, CA, 2008.
58. Hamid, Md.A., Alam, M.M., Hong, C.S. Design of a QoS-aware routing mechanism for wireless multimedia sensor networks. In: *IEEE Global Telecommunications Conference, 2008, IEEE GLOBECOM 2008*, pp. 1–6. IEEE, New Orleans, LA, 2008.
59. Maimour, M. Maximally radio-disjoint multipath routing for wireless multimedia sensor networks. In: *Proceedings of the 4th ACM Workshop on Wireless Multimedia Networking and Performance Modeling*, pp. 26–31. ACM, Vancouver, Canada, 2008.
60. Medjiah, S., Ahmed, T., Asgari, A.H. Streaming multimedia over WMSNs: An online multipath routing protocol. *International Journal of Sensor Networks* 11(1), 10–21, 2012.
61. Stankovic, J.A., Abdelzaher, T.F., Lu, C., Sha, L., Hou, J.C. Real-time communication and coordination in embedded sensor networks. *Proceedings of the IEEE* 91(7), 1002–1022, 2003.
62. Demirkol, I., Ersoy, C., Alagoz, F. MAC protocols for wireless sensor networks: A survey. *IEEE Communications Magazine* 44(4), 115–121, 2006.
63. Cormio, C., Chowdhury, K.R. A survey on MAC protocols for cognitive radio networks. *Ad Hoc Networks* 7(7), 1315–1329, 2009.
64. De Domenico, A., Strinati, E.C., Di Benedetto, M. A survey on MAC strategies for cognitive radio networks. *IEEE Communications Surveys & Tutorials* 14(1), 21–44, 2012.
65. Ye, W., Heidemann, J. Medium access control in wireless sensor networks. *Wireless Sensor Networks*, Springer, pp. 73–91, 2004.
66. Pei, G., Chien, C. Low power TDMA in large wireless sensor networks. In: *IEEE MILCOM 2001*, volume 1, pp. 347–351. IEEE, Washington, DC, 2001.
67. Sohrabi, K., Pottie, G.J. Performance of a novel self-organization protocol for wireless ad-hoc sensor networks. In: *Vehicular Technology Conference, 1999. VTC 1999-Fall. IEEE VTS 50th*, volume 2, pp. 1222–1226. IEEE, Amsterdam, Holland, 1999.

68. Karn, P. MACA—A new channel access method for packet radio. In: *ARRL/CRRL Amateur Radio 9th Computer Networking Conference*, volume 140, pp. 134–140, 1990.
69. Bharghavan, V., Demers, A., Shenker, S., Zhang, L. MACAW: A media access protocol for wireless LAN's. In: *ACM SIGCOMM Computer Communication Review*, volume 24, pp. 212–225. ACM, London, UK, 1994.
70. Lin, C.R., Gerla, M. Real-time support in multihop wireless networks. *Wireless Networks* 5(2), 125–135, 1999.
71. IEEE 802 LAN/MAN Standards Committee et al. Wireless LAN medium access control (MAC) and physical layer (PHY) specifications. *IEEE Standard*, 802(11), 1999.
72. Rhee, I., Warrier, A., Aia, M., Min, J., Sichitiu, M.L. Z-MAC: A hybrid MAC for wireless sensor networks. *IEEE/ACM Transactions on Networking (TON)* 16(3), 511–524, 2008.
73. Ma, L., Han, X., Shen, C.C. Dynamic open spectrum sharing MAC protocol for wireless ad hoc networks. In: *1st IEEE International Symposium on New Frontiers in Dynamic Spectrum Access Networks, DySPAN 2005*, pp. 203–213. IEEE, Baltimore, MD, 2005.
74. Timmers, M., Pollin, S., Dejonghe, A., Van der Perre, L., Catthoor, F. A distributed multichannel MAC protocol for multihop cognitive radio networks. *IEEE Transactions on Vehicular Technology* 59(1), 446–459, 2010.
75. Hu, W., Willkomm, D., Abusubaih, M., Gross, J., Vlantis, G., Gerla, M., Wolisz, A. Cognitive radios for dynamic spectrum access—Dynamic frequency hopping communities for efficient IEEE 802.22 operation. *IEEE Communications Magazine* 45(5), 80–87, 2007.
76. Cordeiro, C., Challapali, K. C-MAC: A cognitive MAC protocol for multi-channel wireless networks. In: *2nd IEEE International Symposium on New Frontiers in Dynamic Spectrum Access Networks*, pp. 147–157. IEEE, Dublin, Ireland, 2007.
77. Hamdaoui, B., Shin, K.G. OS-MAC: An efficient MAC protocol for spectrum-agile wireless networks. *IEEE Transactions on Mobile Computing* 7(8), 915–930, 2008.
78. Cai, L.X., Liu, Y., Shen, X., Mark, J.W., Zhao, D. Distributed QoS-aware MAC for multimedia over cognitive radio networks. In: *IEEE Global Telecommunications Conference (GLOBECOM 2010)*, pp. 1–5. IEEE, Miami, FL, 2010.

4 Multimedia Streaming in Wireless Multimedia Sensor Networks

AHMED HUSSEIN ABBAS SALEM AND RANIA AHMED ABUL-SEOUD

Contents

4.1 Introduction

Multimedia streaming is the process of sending and delivering multimedia content to an end user or to the base station, where it will pass through further processing or be stored for further research. The choice of delivery method is driven by the content that is being distributed. In the case of telecommunications networks, the system uses streaming technology to deliver multimedia content, which might be a video, picture, or audio. Wireless multimedia sensor networks (WMSNs) are a special case where power is critical, as the nodes are mostly battery-operated; advances in the protocols and methods of streaming are crucial in this case. For video, there is a recommended broadband speed; if this recommendation is satisfied, the video content will be meaningful and useful. The minimum recommended broadband for video is 2.5 MB/s, which consumes a great deal of power, particularly if the network is used for real-time monitoring and surveillance. For high definition video, the minimum recommended broadband is 10 MB/s. As a result, compressing the multimedia content is crucial; audio and video codecs are used to compress the data as much as possible without harming it or causing it to be meaningless. Compression conserves power and increases efficiency in many applications, such as monitoring nearby areas. However, high resolution with raw, unmodified content can be useful, especially if the field of view is broad; the user can then magnify an area of the picture without distortion. This functionality is advantageous if the area being monitored is very large, such as a whole town; although in this case a very expensive camera would be required.

Particular interest has been shown in developing an advanced multimedia streaming method to prevent the occurrence of dead nodes, which is especially important if the nodes cooperatively send content to the base station using routing protocols. However, such situations can be solved by choosing the best-fitting MAC protocols, routing protocols, network topology, and method of power supply. Nodes can be powered using solar cells, although this method has drawbacks because the cells need to be directed toward the sun; solar power is not always guaranteed, especially in harsh environments like forests. Developing streaming techniques differs from computer networks. A computer network works toward a primary issue, to prevent delays, but WMSNs work toward a different primary issue—to prevent high power consumption, which may lead to dead nodes in the network. These problems will be solved by using the smart advanced multimedia streaming techniques that will be shown in this chapter.

4.2 Differences between Wireless Sensor Network and WMSN Requirements in Multimedia Streaming

WSNs measure scalar data such as temperature, humidity, and other data that can be represented as a measured value. These data are measured by a sensor and is then sampled, calibrated, packetized, and sent to the base station. This process and its requirements are very different from those in WMSNs, which primarily send multimedia content. The hardware differences are very clear; scalar data require only a sensor and an integrated circuit (IC) that has an analog to digital channel (ADC) channel. By contrast, multimedia requires a camera, microphone, and hardware that can provide the required and needed processing for such data. The differences between a WSN and a WMSN are as follows:

- Quality-of-service requirements: Multimedia applications need different requirements to supply different functionalities, such as snapshots, playback, and storage. Different functionalities will require specific limits on the delays encountered and the jitter.
- Bandwidth: WMSNs require several times more bandwidth than was needed to support ordinary scalar sensor networks.
- Power: Power consumption is higher for WMSNs, because it is consumed by data storage activities, high transmission rates, and the aggressive processing applied to multimedia content.
- In-network processing support: WMSNs have the availability and the support to extract targeted data from the content stored, which could be used in object identification applications, and the usage of image fusion capabilities will enhance the capabilities of such application.
- In-node processing support: Such support is needed to compress the raw files stored and supplied from the cameras and the microphones in WMSNs.
- Cross-layer design: Complete implementation and optimization is necessary in WMSNs from the application layer to the physical layer for the node to work and cooperate in the network without encountering any incompatibility or lack of support issues.
- Better hardware design: WMSN design has been concerned with managing power consumption in a better way to conserve energy, which has accelerated the need to investigate energy-harvesting methods.

4.3 Multimedia Streaming Optimization Techniques

Optimization techniques are a necessity in WMSNs to increase the efficiency of data transfer. Some of these techniques are as follows:

- Deduplication: Redundancy data will be eliminated by using references instead of the actual data by working on the bytes of the data. The benefits of such a technique will be obvious in IP applications.

- Compression: Compression in WMSNs is similar to ordinary compression programs as it relies on data patterns that can be represented in an efficient manner. It is very important to observe the difference between compression technique and compressive sensing. Compressive sensing is also known as compressive sampling and sparse sampling. It is an efficient way to acquire the signal and reconstruct it by finding solutions to some underdetermined linear systems that use the sparseness and the compressibility of the signal, which will extract the entire desired signal with few measurements.
- Latency optimization: Latency optimization is accomplished by giving the authority of answering and replying to different requests from the nodes in a local manner instead of sending such requests to remote administrative nodes to reply.
- Forward error correction: Its main goal is to reduce retransmissions, which occur especially in harsh environments or bad weather, by adding a loss recovery packet for every specific number of packets being sent.
- Connection limits: Connection limits represent a security method to ensure the avoidance of denial of service in WMSNs.
- Simple rate limits: Congestion can be avoided by limiting the rate at which packets are sent to any user.
- Multimedia in-network processing: Multimedia processing was mostly dealt with as a separate problem from network design. Studies have primarily taken an interest in the cross-layer interactions among the lower layers of the protocol stack. However, the processing and streaming of multimedia content are not separate or independent from each other, and their function has a great impact on the quality of service delivered to the user and required by the application, as well. Processing is crucial in the case of WMSNs and, in point of fact, in any video streaming application. Video is first recorded as a raw file, which is a digital or analog record that has been digitized without any occurrence of compression or distortion to the file itself. Using this type of file will consume a great deal of storage and processing as well as considerable power, which is a big disadvantage

for WMSNs. Thus, files must be compressed to be ready for streaming. Compression may decrease the quality of the files and make low-quality videos worse; audio files will be hard to hear or understand. Therefore, methods for reducing file size are similar to taking smaller pictures or reducing the frame rate. Reducing the frame rate would be useful in many applications, such as monitoring melting polar ice. For this application, it would suffice to record video at specific predefined intervals for short periods, or just to take a picture. There are many advances necessary to make such an application possible. Problems include the power, the supports for the camera, and how to keep the camera lens and microphone clean.

4.4 Key Video Parameters

The key video parameters must be known and identified to determine the quality of the received multimedia content through the wireless channels. The parameters are as follows:

- Sample quantization rate: This parameter is the number of bits being used to quantize each sample, measured in bits/sample unit. The lower the number of bits used, the lower the amount of information sent for each sample. This will allow a greater number of samples to be sent, which is advantageous. However, a greater quantization error will be encountered in each sample. The benefit of sending more samples is to outweigh the distortion encountered in the samples; a minimum rate would be 5 bits/sample. Moreover, image corruption is acceptable, as a corrupted image might appear once in every 103 sent images.
- Samples per frame: This parameter is the required number of samples that must be supplied to reconstruct an image with a predefined level of quality. Clearly, the quality of the image will increase with an increase in the number of samples sent. All this will depend on the required quality of the video.
- Channel encoding rate: This could be identified by determining the channel coding strategy that is best suited for transmitting packets over a multihop wireless network.

4.5 Operations on Multimedia Content

This section will illustrate the operations that must be performed on multimedia content to prepare it for packetizing and sending. Raw files cannot be sent in their original form; some operations must be performed on them to make them ready and to conserve the power, processing, storage, and bandwidth usage. There are many operations that could be done; we will illustrate some of the most popular methods in the WMSN and multimedia arenas. Much of the recent research discusses compression, which is found in almost every multimedia application. The JPEG is a common image compression algorithm that is found in most CMOS camera sensor chips, as JPEG can achieve a compression ratio from 10:1 to 20:1 without a noticeable loss in color images. A compression ratio from 30:1 to 50:1 is also possible with small or moderate distortion in the image. Compression of JPEG images is mostly lossy, and lossy compression means that some of the image content will be reduced. In color images, the raw file is divided into two components: luminance and color components. The difference here is only in the representation of the data; no loss or reduction have been achieved. The human observer is much more observant of the intensity information than the color information itself. Therefore, the color information can be subsampled without any noticeable defect in the image, although the amount of data will be highly reduced. The image is then divided into 8 × 8 pixel blocks and the algorithm is performed on these blocks. The blocks will be transformed by using the discrete cosines transformation method. The next step is the quantization step, which is crucial for the quality that the user needs. As the quality increases, the size of the image will increase, which might require more storage, power, processing, and bandwidth. The JPEG is a block-based compression method. The compressed sensing (CS) method is also common and is heavily used, especially in WMSNs; the CS method is more flexible than JPEG and consumes less power. The CS method has several advantages—for example, an inherent resilience to channel errors, which are caused by unstructured image representation; this will lead to a zero loss in image quality. In the CS method, the transmitted samples constitute a random and incoherent combination of the original data itself. No single sample is given priority over any other sample, and only the number of completely received samples that have not encountered any errors are used to construct the image. Thus, the complete samples are the main parameters for achieving a certain quality to the image; clearly, the image quality will increase when the corrupted samples are dropped. The corrupted samples are dropped in adaptive parity-based channel coding, which was an advance introduced to the CS method. In adaptive parity-based channel coding, the quality of the video stream can be improved by dropping the samples that are corrupted or have encountered some errors that might impair image reconstruction by introducing incorrect information. Adaptive parity-based channel coding can be achieved by using even parity on a definable number of samples that will be dropped at the receiver or at an intermediate node in a multihop scenario if the parity check has failed. Adaptive parity was tested against rate compatible punctured convolutional (RCPC) codes; it outperformed it at all levels of the RCPC codes. The adaptive parity method has performed better in all the bit error rate (BER) levels. BER is the number of bits received by a receiver through a wireless channel that has encountered noise, interference, and distortion errors. Adaptive parity performs better. Although forward error correction (FEC) schemes have more powerful error correction methods, FEC has no additional overhead. This additional overhead is needed to increase video quality, but instead FEC drops the samples that have encountered errors. There are several error correction schemes, codes, and methods, but the parameters that were targeted while designing them were the capability to correct any kind of error, the time needed to correct such errors, and sufficient memory to power the algorithms. These systems were created for computers, where power was not a large problem because the processor, RAM, and storage cooperated with the interfaces and peripherals of the system. However, in WMSNs, power is a great concern, so data must be sent in a lossy manner. In WMSNs, the algorithms depend on dropping the error rather than retransmitting the packet to extract the correct information. Future development in WMSNs should target reliable, fast, and energy-efficient methods.

4.6 Conclusion

It was shown that, although JPEG encoded images are used extensively, they are not best suited for WMSNs. The CS method will perform much better because of its inherent resiliency to channel errors.

It was also shown that FEC methods and schemes are not best suited or beneficial for WMSN applications. The adaptive parity method will perform much better because it is able to drop samples that have encountered errors, thus improving the quality of the multimedia content that was received, while maintaining low complexity.

5

Coverage Problems for Wireless Multimedia Sensor Networks

XIAOLAN LIU, GUILIN CHEN, AND BIN YANG

Contents

Recently, the rapid development and progress of low complexity, high flexibility sensor facilities (visual sensor nodes, digital signal processing devices, etc.) has encouraged the emergence of wireless multimedia sensor networks (WMSNs). Multimedia sensor nodes are capable of gathering richer multimedia information (besides scalar physical phenomena) from a monitored region. Hence, WMSNs have been widely applied to various scenarios, such as smart homes, medical applications, and so on. Recently, coverage as a measure of the quality of service (QoS) has drawn much attention in WMSNs. Compared to traditional sensor networks, multimedia sensor networks have exclusive properties, such as sensor viewing direction and angle, as well as object facing direction and location. Based on the above considerations, the coverage problems in WMSNs require specific algorithms and solutions. In this chapter, we mainly survey the existing research in the field of multimedia device coverage. First, we list and classify off-the-shelf sensor hardware, as well as available coverage models. Furthermore, we describe the state of the art in influencing factors, deployment mechanisms, sensor selections, and performance metrics for WMSN node coverage. The coverage problem in WMSNs can be divided into three categories: barrier coverage, area coverage, and target coverage. We discuss the three types of coverage and note representative solution algorithms. Finally, we discuss the existing problems and new research trends in key areas. We conclude that WMSN coverage research is still an open issue that has not been fully addressed.

5.1 Introduction

The recent development of low complexity, high flexibility sensor devices, such as audio and visual sensors, has motivated the emergence of WMSNs, that is, a system of interconnected multimedia sensor motes that consists of a sensing unit, processing unit, communication unit, memory units, power unit, coordination unit, and so on. WMSNs can obtain richer multimedia information, as well as scalar physical phenomena, from a monitored site. In addition, they are also capable of storing, processing, transferring, and fusing multimedia content originating from heterogeneous sensor devices [1]. Hence, WMSNs promise a number of prospective applications. Most of the key applications are divided into surveillance, management, monitoring, detection, object tracking, and virtual reality.

Coverage, which evaluates how well a target area is monitored, is a fundamental performance metric to measure QoS [2]. It has drawn researchers' attention recently, leading to several papers about coverage. For the most part, research has focused on traditional sensor coverage [3–6]. However, the coverage problem in WMSNs has drawn a great deal of attention from the research community in recent years. Ma and Liu [7] present the concept of WMSNs and discuss multimedia sensor coverage problems. Compared to traditional sensor nodes, multimedia sensor nodes have exclusive properties, such as sensor viewing direction and angle, as well as object facing direction and location [8]. A comprehensive discussion of the traditional coverage problems in WMSNs is provided in Ref. [9].

Because most of the existing coverage algorithms are confined to the field of traditional sensor networks, they are not a good fit for multimedia sensor networks. The coverage problem requires specific solutions and techniques in WMSNs. In this chapter, we mainly survey the existing research in the field of multimedia node coverage, including hardware devices, coverage models, affecting factors, deployment mechanisms, sensor management, performance metrics, and typical coverage, and discuss future research problems in detail.

In this chapter, we further divide the coverage content into four stages: design stage, deployment stage, management stage, and coverage metrics. For the design stage, we list some relevant factors to determine how many multimedia nodes are needed to provide complete field coverage in the monitored environment. For the deployment stage, we divide the deployment type into two groups: random placement and deterministic placement. Random placement, where multimedia nodes may be scattered via a plane or launched by artillery, is mainly adapted to inhospitable or remote regions. Multimedia nodes may also be placed in a predetermined position in a deterministic way. In order to satisfy the specified application requirements, sensor management mainly applies to select nodes and schedule the activity of different nodes, while maintaining multimedia node coverage. Finally, the coverage metric which includes network connectivity, sensor lifetime, and node energy is an indicator of the quality of coverage.

In this chapter, the reader may gain better understanding of current coverage research in the field of WMSNs. Simultaneously, it may also promote discussion and encourage new research topics within the research community.

The reminder of the chapter is organized as follows. In Section 5.2, we list real-world multimedia node devices and present their off-the-shelf hardware platforms as well as architecture. In Section 5.3, we list and classify available coverage models, such as mathematical model, physical model, etc. In Section 5.4, we introduce the key factors affecting multimedia sensor coverage. In Section 5.5, we discuss multimedia sensor deployment and possible scheduling plans. In Section 5.6, we investigate the existing research in sensor management. Section 5.7 presents several coverage performance metrics. The coverage type in WMSNs can be classified into three categories: barrier coverage, target coverage, and area coverage. In Section 5.8, we describe several different types of coverage and review representative solution algorithms in the chapter. Open research issues are discussed briefly in Section 5.9. Section 5.10 concludes the chapter.

5.2 WMSN Hardware

In this section, we first introduce and classify existing sensor nodes that can satisfy the special applications in WMSNs. Additionally, we describe multimedia node structure with a particular emphasis on multimedia node sensing motes. Moreover, we show existing hardware platforms and test beds. Eventually, we review existing wireless multimedia network architecture which contains homogeneous and heterogeneous.

5.2.1 Multimedia Sensors

In addition to traditional sensor nodes such as temperature sensors, pressure sensors, and humidity sensors, WMSN sensors also include some other multimedia sensors. Figure 5.1 shows several real-world multimedia sensors, including video sensors, infrared (IR) sensors, and ultrasound sensors. In the following subsection, we introduce several multimedia sensor nodes and describe their unique performance.

5.2.1.1 Video Sensors Video sensors, in the form of numerous low-cost complementary metal–oxide-semiconductor (CMOS) camera nodes, have brought new opportunity for applications within the scope of visual surveillance and traffic security monitoring. In these applications, video sensors retrieve information-rich visual data from a monitored physical environment.

With the development of high-resolution visual technology, the large-scale products of video sensor nodes have been used in WMSNs, which contain robots, toys, computers, cell phones, etc.

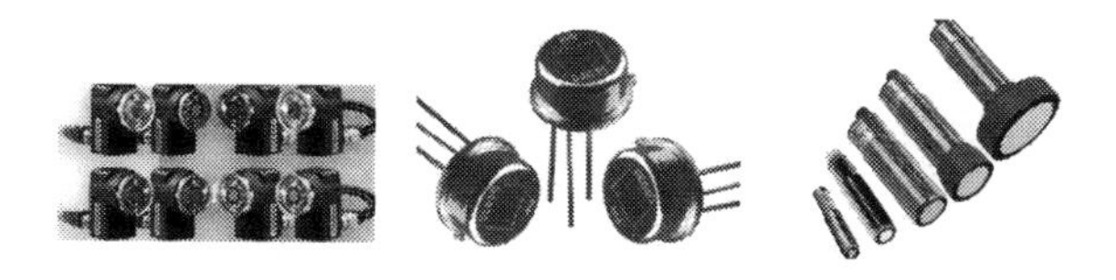

Figure 5.1 Multimedia sensors, from left to right: video sensors, infrared sensors, and ultrasound sensors.

Some related research is found in Ref. [10], where Akyildiz et al. present an overview of WMSNs, including networking architectures, communication protocol, and multimedia sensor hardware. To discuss video sensor nodes robustly, Ref. [11] consider future research directions in the field, including sensor hardware processing, communication, and sensor management.

To illustrate the physical properties of video sensors, we first introduce some camera terminology [12]: depth of field (DoF), focal length (f), focus distance (s), hyperfocal distance (H), the angle of view (AoV) [13], and so on. In the following introduction, we focus on AoV. As the maximum visible volume, AoV can be measured vertically, horizontally, and diagonally. In some literature, the term field of view (FoV) denotes AoV [13]. But the FoV differs slightly from AoV, as is proven by Ref. [14], which describes the sensing radius as equal to the AoV of a 1/3" lens for a video sensor.

Recently, with the improvement of image capturing devices, charge-coupled device webcam, and the CMOS imaging device [15], video sensors have been produced extensively. Most video sensors are manufactured by Agilent [16] and Omnivision [17]; Table 5.1 provides a description of the features of several video sensor products (e.g., ADCM-1700, 2650, 2700 vs. OV6620, 7620, 9630).

5.2.1.2 IR Sensors IR sensors are used to monitor specific properties of the physical environment according to emit and detect IR radiation. They also have the ability to detect a target's heat and motion [18]. We classify IR sensors into several types. One type is passive IR (PIR) sensors, which are mainly used to detect IR rays, rather than emit them [13].

Table 5.1 Comparison of Several Video Sensors

VIDEO SENSOR	COMPANY	PLATFORM	IMAGE TECHNOLOGY	LENS SIZE	f-NUMBER	DEFAULT RESOLUTION
ADCM-1700	Agilent	MeshEye	CMOS	N/A	2.8	352 × 288
ADCM-2650	Agilent	N/A	CMOS	N/A	2.8	480 × 640
ADCM-2700	Agilent	MeshEye	CMOS	N/A	N/A	640 × 480
OV6620	Omnivision	CMUCam3	CMOS	1/4	N/A	352 × 288
OV7620	Omnivision	CMUCam3	CMOS	1/3	N/A	640 × 480
OV9630	Omnivision	N/A	CMOS	1/3	N/A	1280 × 1024

Source: Amac Guvensan, M., and Gokhan Yavuz, A., *Ad Hoc Networks*, 9, 1238–1255, 2011.
Note: N/A, not available.

This ray is not visible to the human eye, but it is extremely vulnerable to be interpreted by other IR radiation [18].

A special IR sensor such as a motion sensor can use optics or acoustics to detect an object's behavior. Such behavior might be either active or passive. Active IR sensors detect objects' behavior by measuring the feedback of optics or sound waves. In addition, PIR sensors are typically used to monitor indoor environments and linked to IR burglary protection systems [19]. There exist many other IR sensors (e.g., reflective IR sensors and interrupter IR sensors); it is impossible to introduce all IR sensors individually.

The working principle for IR sensors is as follows. First, the IR sensor detects IR rays. The next step is to transform IR rays into electric current. Finally, we detect motion using an amperage or voltage detector [20]. Compared to other radiation, the wavelengths of IR radiation are longer than visible ray wavelengths in the electromagnetic spectrum and shorter than microwaves [21]. We often find IR radiation in daily life. For instance, there is an IR detector in every television, which can analyze the signal from a remote base station. All in all, IR sensors have been used in a wide range of applications because of their unique features—low power, portability, and simplicity [21].

5.2.1.3 Ultrasound Sensors An ultrasonic sensor is a device that uses high-frequency sound to measure distance [22]. The basic operating principle for ultrasonic sensors is as follows. The ultrasonic sensor launches a sonic pulse, and then the pulse bounces off an object. The transducer converts among electrical, sonic, and mechanical energies [23].

In addition, ultrasonic sensors have many applications. For example, medical facilities often use ultrasonic sensors to visualize areas within the human body. Industrial machines often use these sensors to detect the presence of a living being in an automated factory. Security installations also use sonic sensors to detect the presence of an unauthorized person [22]. However, the drawbacks of ultrasonic sensors are that we cannot tell the difference between large bodies and small bodies. The reason is that the emitted sonic pulse is cone-shaped. Ref. [23] proposes to use multiple sensors or rotating sensors to solve this problem.

5.2.2 Node Structure

The structure of a WSMN node is designed to be simple, small, and low cost. The typical hardware components of a multimedia sensor device include a sensing unit, central processing unit (CPU), memory unit, communication unit, and power supply unit, and so on, as shown in Figure 5.2. We will illustrate the major component motes of WMSN nodes.

- The sensing unit contains several different multimedia sensors and analog-to-digital converters (ADCs). In the sensing stage, the multimedia sensors collect the analog signals from a monitored environment. Then the analog signals are converted into digital signals through the ADC. Finally, the digital signals are delivered into the processing unit.
- The processing unit is often designed for a specific application such as digital signal processors, because of better performance and lower power consumption. In addition to regulate of communication, coordination, network synchronization of the unit, and so on, the most important assignment of the

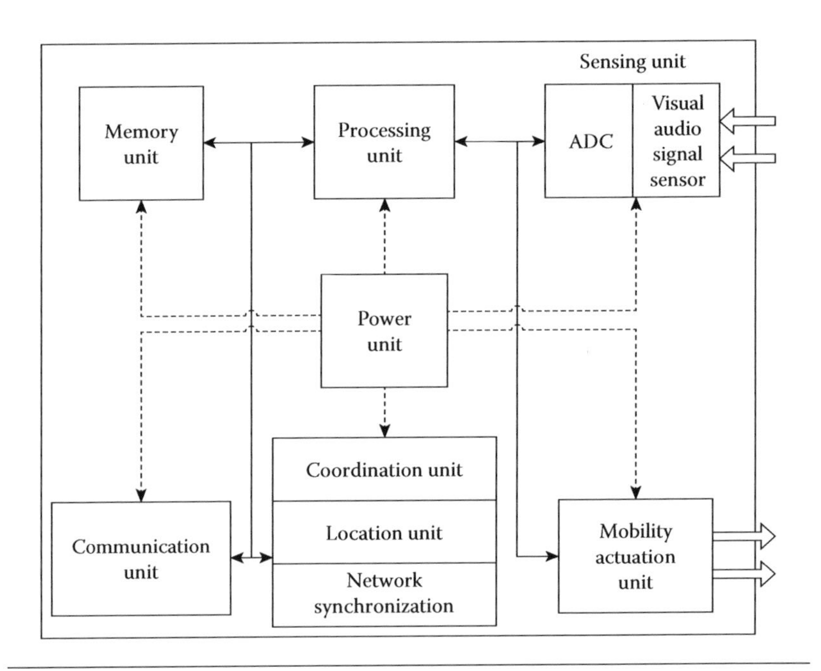

Figure 5.2 Basic components of the multimedia node.

processing unit is to process the data from the sensing unit. Afterwards, the processed data are conveyed to the base station or the memory unit.
- Generally speaking, the memory unit is used for storing the sensed data or procedure code. Once in a while, it functions as a multimedia data buffer. Hence, there is a high storage requirement for WMSNs, as WMSNs handle a large amount of multimedia data, such as images, audio, video, as well as scalar sensor information. In total, the storage requirements for WMSNs are at least ten times higher than scalar networks.
- Currently, the communication protocols are mainly based on the Institute of Electrical and Electronic Engineers (IEEE) 802.14.3 or ZigBee standard [10]. The coordination unit is responsible for regulating location and network synchronization. The mobility/actuation unit can move or manipulate targets.

Most multimedia devices are powered by portable batteries that cannot be recharged or replaced without delay. Therefore, energy efficiency has become an important research topic for WMSNs. However, all research on battery energy efficiency has been for scalar sensors. The content for multimedia sensors has yet to be explored.

5.2.3 Hardware Platforms and Test Beds

To handle more multimedia node applications and to test and examine the above-mentioned algorithms and protocols for WMSNs, we survey some existing hardware platforms and test beds. We examine multimedia data transmission efficiency in terms of bandwidth, storage, data rate, power consumption, and so on [24]. In this subsection, we introduce available hardware prototypes and categorize the existing platforms and test beds based on their capabilities and functionalities. Figure 5.3 shows the classification for most off-the-shelf platforms and test beds.

5.2.3.1 Hardware Platforms Existing multimedia platforms and research prototypes are shown in Table 5.2, which displays the different capabilities and features of hardware platforms. In terms of node structures such as the processing unit, memory unit, power unit, sensing unit, and radio, we classify these hardware platforms

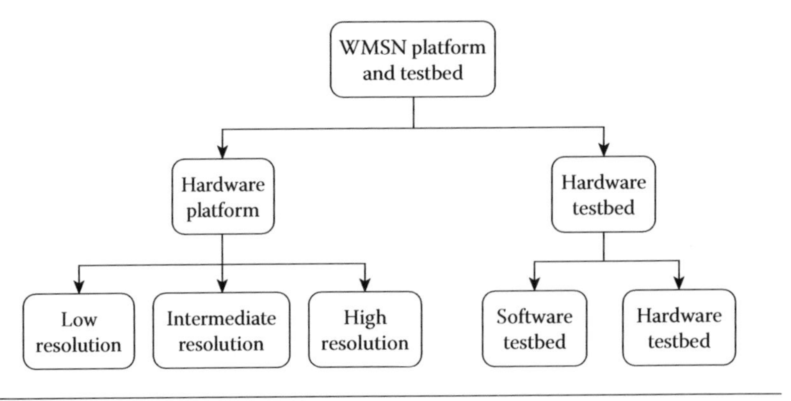

Figure 5.3 Wireless multimedia sensor network platforms and test beds.

Table 5.2 Comparisons of Hardware Platforms for Wireless Multimedia Sensor Networks

PLATFORM	IMAGE SENSOR	PROCESSING UNIT	MEMORY	RADIO	ENERGY (mW)
Cyclops [25]	CMOS	8-bit ATMEL ATmega128L MCU + CPLD	512 KB flash 64 KB SRAM	ZigBee	110–0.76
CMUcam3 [26]	CMOS	ARM7TDMI (32-bit) 60 MHz	64 KB RAM 128 KB flash	ZigBee	572.3–0.29
Stargate [27]	Webcam	PXA255 XScale 400 MHz	32 MB flash 64 MB SDRAM	IEEE 802.11 Bluetooth	N/A
Imote2 [28]	Webcam	32-bit PXA271 XScale	256 KB SRAM, 32 MB flash 32 MB SDRAM	ZigBee	322–1.8
Panoptes [30]	Webcam	400 MHz 32-bit PXA255 XScale CPU (Stargate)	32 MB flash 64 MB SDRAM	IEEE 802.11	5300–58
MeshEye [31]	CMOS	55 Hz 32-bit ARM7 TDMI ATMEL AT91SAM7S	64 KB SRAM, 256 KB flash	ZigBee	175.9–1.8

Note: N/A, not available.

into three groups: low-resolution hardware platforms, intermediate-resolution hardware platforms, and high-resolution hardware platforms.

- Low-resolution hardware platforms are the low-power processing devices platforms; they were specifically designed for simple object detection task from multiple disparate viewpoints. Cyclops, Cyclops Electronics Company, Glasgow [25] and CMUCam3, Carnegie Mell University, Pennsylvania [26] are examples of low-resolution hardware platforms.

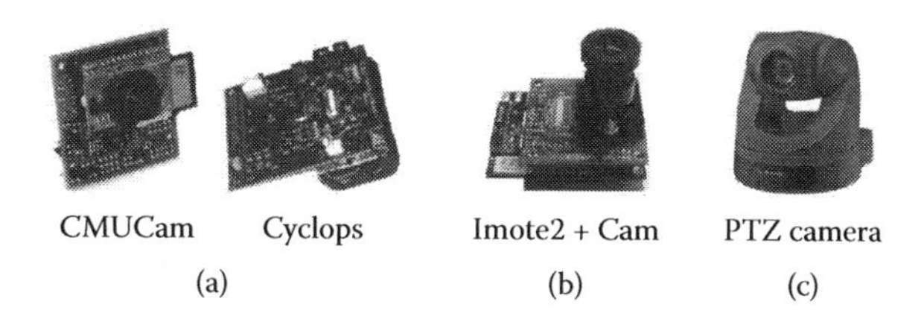

Figure 5.4 Examples of wireless multimedia sensor network hardware platforms. (a) low-resolution camera; (b) intermediate-resolution camera; and (c) high-resolution camera. (From Almalkawi, I.T. et al., *Sensors*, 10, 6662–6717, 2010.)

- Intermediate and high-resolution hardware platforms are designed for sophisticated high-quality applications, such as network management, target detection, and monitoring. These two kinds of hardware platforms can consume a great deal of energy. As the examples of intermediate resolution hardware platform, the Stargate board [27] and the Imote2 [28] platforms (Intel company, California) and beyond that PTZ camera [29] (Sony Corporation, Tokyo, Japan), Panoptes camera [30] (Logitech company, California), and MeshEye camera [31] (Agilent Technologies incorporated Corporation, California) are considered as high-resolution cameras. Figure 5.4 shows several commercial products of hardware platforms used in WMSNs.

5.2.3.2 Test Beds To evaluate different processing algorithms and communication protocols or test various applications in WMSNs, researchers obtain experiment results and theoretical analysis according to conduct simulations, because the experimental formula in WMSNs is inefficient and complex, and it is difficult to be repeatedly used by other researchers. As a result, simulation reduction is considered the best measurement methodology in WMSNs. However, current simulators cannot model the major characteristics of real-time multimedia systems. Obviously, simulation reduction is somewhat dubious and has limited credibility [32]. In order to remedy the gap between theoretical and actual approaches, researchers have evaluated proposed protocols and algorithms in a test bed.

According to whether they can test and evaluate application, protocols, research prototypes, or network performance metrics in real environments, test beds can be divided into two categories: hardware test beds and software test beds [24]. Software test beds are designed on application program interfaces, which provide testing and evaluation of application conditions via abstraction layers that hide the low-level hardware devices. WiSNAP [28,33] is an example of a software test bed. Hardware test beds consist of some hardware devices, such as multiple sensor nodes, wireless communication hardware, base stations, supporting tools for user interface and information monitoring.

According to the hierarchal organization in WMSN, we further divided hardware test beds into single-tier and multitier test beds. A single-tier hardware test bed, Meerkats [34] is used for detecting and monitoring wide surroundings. IrisNet [35], a multitier hardware test bed, builds large-scale distributed networks for heterogeneous WMSN applications. Table 5.3 gives a summary of the features for existing software and hardware test beds.

5.2.4 Network Architecture

Similar to traditional sensor networks, the network architecture of WMSNs is also a flat, homogeneous architecture, where each multimedia node device has a similar or identical processing capability. However, homogeneous architecture is not suited for processing multimedia applications. We introduce various heterogeneous architectures for WMSNs in Figure 5.5, where the network architecture of WMSNs can be divided into three categories in accordance with different network application characteristics.

- Single-tier flat architecture consists of homogeneous multimedia nodes. They have the same processing capabilities in the network architecture, where all the nodes have the ability to accomplish any function to the sink by means of multihop route, as shown in Figure 5.5a.
- Figure 5.5b shows a single-tier clustered deployment architecture, which contains heterogeneous multimedia nodes, such as visual, image, scalar sensors, and so on. They pass on the sensed information to the cluster head, which has a more complex ability to process data. The cluster head is joined with the gateway or the sink either directly or through other cluster heads through a multihop route [24].

Table 5.3 WMSN Test Beds

	TEST BED NAME	CAMERA AND RESOLUTION	WIRELESS MOTE	ADDITIONAL FEATURES
Software Test Beds	WiSNAP	Includes device library Agilent ADCM-1670	Includes device library of Chipcon CC2420DB IEEE 802.15.4	-MATLAB®-based test bed -Open source APIs -Multimedia processing Primitives
	AER Emulator	OmniVision OV7649 640 × 480 @ 30fps 320 × 240 @ 60fps	XYZ, Imote2 IEEE 802.15.4	-Visual C++ based test bed -AE recognition
Hardware Test Beds	Meetkat	Logitech QuickCam Pro 4000 640 × 480	Stargate IEEE 802.11b	-Energy efficient -Event detection
	SenseEye	Cyclops, CMUcam3, PTZ Sony SNC-RZ30N Different resolutions	MICA2 IEEE 802.15.4 Stargate IEEE 802.11	-Multilevel resolution -Surveillance application
	IrisNet	Logitech QuickCam Pro 4000 640 × 480	Stargate	-Internet-like queries -Scalable
	Explorebots	X10 Cam2 320 × 240	MICA2 IEEE 802.15.4	-Mobile robot -Electronic compass Ranging devices for navigation
	Mobile Emulab	Overhead Hitachi KP-D20A 768 × 494	MICA2 IEEE 802.15.4 Stargate IEEE 802.11b	-Mobile robot -Evaluate mobility-related network protocols
	WMSN test bed	Logitech QuickCam Pro 4000 640 × 480 176 × 144 @15fpx	MICAz IEEE 802.15.4 Stargate IEEE 802.11b	-Mobile robot -Multilevel resolution

Source: Almalkawi, I.T. et al., *Sensors*, 10, 6662–6717, 2010.

- The middle section presents a multitiered network, with heterogeneous multimedia sensors. In the architecture, the low section is composed of tradition sensors, which executes simple tasks like object sensing. The middle section deployed with visual nodes may execute more complicated tasks such as target capturing. The upper section, with high-resolution multimedia nodes, has the ability to execute more complex tasks, like target tracking. Any tier has a central hub for processing data and exchanging information with the upper tier. The final tier is used for linking the gateway or the sink [24], as illustrated in Figure 5.5c.

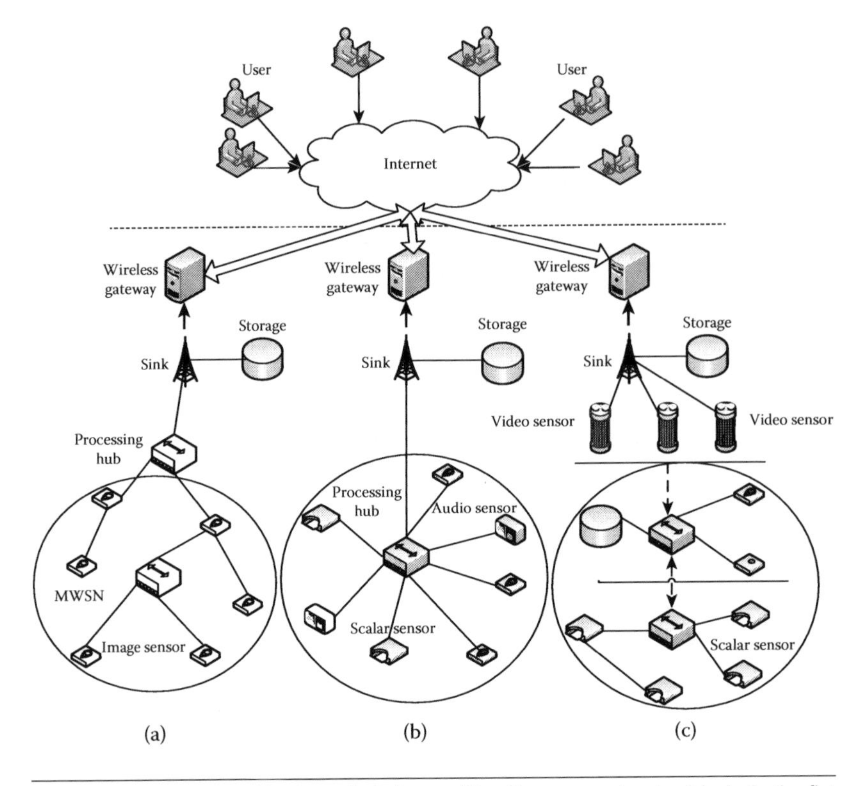

Figure 5.5 Network architecture of wireless multimedia sensor networks: (a) single-tier flat architecture; (b) single-tier clustered architecture; and (c) multitier architecture.

Table 5.4 Hardware Platforms, Typical Tasks for Each Tier in a Multitiered Hierarchical Architecture

HIERARCHY	TYPICAL TASK	HARDWARE PLATFORMS
Low tier	Object sensing	Cylops/XYZ-ALOHA/ MeshEye
Medium tier	Object capturing	MeshEye/Firefly Mosaic/CITRIC
High tier	Object tracking	Panoptes/Meerkats

Source: Tavli, B. et al., *Multimed. Tools Appl.* 60, 689–726, 2012.

Several network architectures have many applications, especially for multitiered heterogeneous network architecture, where each tier sensor is in charge of different functionalities and with different hardware platforms. Table 5.4 presents hardware platforms for each tier in a multitiered hierarchical architecture [36].

5.3 Coverage Models

The coverage problem estimates the capturing capability of multimedia nodes and QoS by obtaining the geometric relation among the target and sensor nodes. Generally speaking, coverage models might be formulated as the function of the angles or the Euclidean distances between the target and the sensor node [37]. There are several attributes for categorizing available coverage models, as shown in Figure 5.6.

In this section, we mainly discuss two coverage models: the mathematical coverage model [38] and physical coverage model. In addition, we review the other off-the-shelf coverage models such as the two-dimensional (2D)/three-dimensional (3D) coverage model and full-coverage model, as well as the full-view coverage model and non-overlapping and overlapping coverage models. In the following, we will provide a detailed introduction to the above-mentioned coverage models.

5.3.1 Mathematical and Physical Coverage Models

Mathematical coverage models depict the geometric relationship of coverage by a multimedia node or node system. To deal with the coverage relation, we divided coverage model into the Boolean (binary) coverage model and the probabilistic coverage model.

In the Boolean coverage model, the coverage measurement is either 0 or 1 to express a target. If an object is located through sensor's capturing scope, the coverage measure is considered as 1. Otherwise, it is 0 and the object will not be perceived due to interference factors, even if the target is located within the sensing area in practical applications.

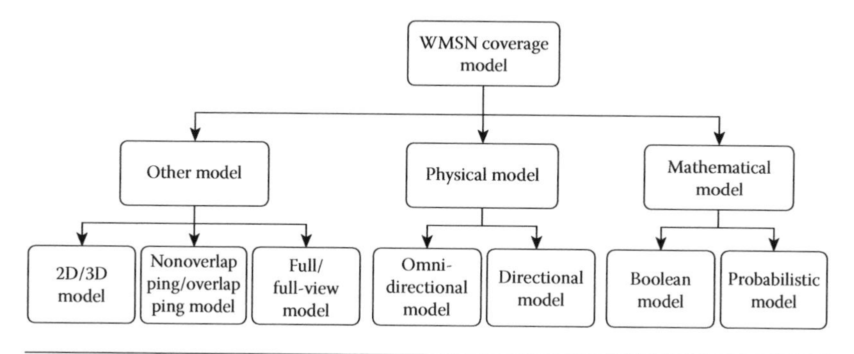

Figure 5.6 Comparisons of coverage models.

As a result, the Boolean coverage model is an ideal model for multimedia sensor coverage.

The probabilistic coverage model [39] describes sensed objects within the scope of the multimedia node via the probability function, which function is described by either the distance or angle between the multimedia nodes and the target, and the value of the probability function decreases as the distance/angle descends.

Physical coverage models can provide sensing direction information for multimedia sensors. We classify physical coverage models into the omnidirectional (isotropic) coverage model and directional coverage model.

The sensing range of the omnidirectional model is often abstracted as a circular area and a target is considered to be captured or covered by a multimedia node if it is in the field of the sensing range for the multimedia node [40]. At present, most of researches mainly use the circular sensing model to solve coverage issues [41–44].

Unlike omni-directional model, directional model can record different views of the target or different direction targets [40]. Directional nodes have a finite AoV and cannot sense the entire circular scope. Consequently, the coverage range of directional sensors is a fan-shaped area in a 2D plane.

According to a simple geometrical abstraction, the sector coverage model of directional nodes is expressed by fourfold $<S_i, r, \varphi, f>$, where S_i expresses the location of the directional node, r is the maximum sensing radius, φ is a FoV angle, and f represents the working orientation of the object. The sensing area of the directional sensor for a 2D plane is shown in Figure 5.7a. A point P is covered by a node S_i if P is within the sensing scope of S_i, as shown in Figure 5.7b.

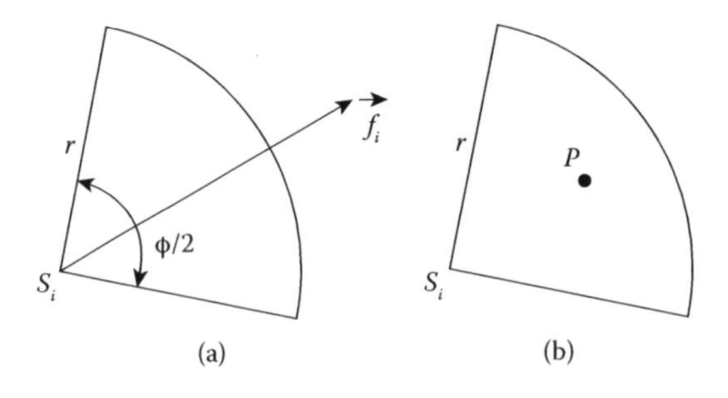

Figure 5.7 (a) The directional sensor S_i sensing sector and (b) an object point covered by node S_i.

5.3.2 *Other Coverage Models*

In addition to the above-mentioned coverage models, there are some other coverage models according to different classifications. For example, coverage models might be divided based on their description of a spatial region into 2D and 3D coverage models. Based on the directional relationship between target and multimedia sensor, coverage model can be divided into full-coverage model and full-view coverage model. We divide coverage models into overlapping and nonoverlapping coverage models according to the relationship between multimedia sensors. In the following subsection, we will introduce several different coverage models in detail.

5.3.2.1 2D/3D Coverage Model To simplify the experimental estimation of the proposed algorithms in the existing publications, most coverage models regard the monitored area as 2D coverage model. However, the 2D coverage model does not meet real-world application requirements. Compared to the 2D model, the 3D model can deliver more accuracy. The establishment of the 3D model mainly focuses on two distinct features for pan–tilt–zoom (PTZ) directional sensors; one is a multimedia sensor with a fixed 3D point. The other is the coverage area as a projecting quadrilateral field in a 2D surface.

The 3D coverage model is represented by five elements (P, W_d, A, α, and β), where P expresses the sensor position (x,y,z), W_d indicates the working direction for the directional node, A is the maximum value of the FoV angle, and α versus β represent the horizontal and vertical offset angles in sight around W_d. $W_d = (dx(t),dy(t),dz(t))$ denotes a unit length, where $dx(t)$, $dy(t)$, and $dz(t)$ are elements along the x, y, and z axes, respectively [13]. Figure 5.8 shows the 3D coverage model [45]. If an object P is covered by a sensor S_i, the following condition must be satisfied: the object P must be fixed in the projecting quadrilateral area in a 2D surface.

5.3.2.2 Full/Full-View Coverage Model The full-coverage model has nothing to do with target direction, if only the object is within the node's sensing range. As to some special applications, multimedia sensors not only record an object image, but they must also obtain a frontal view of the object to identify the target clearly. As a result,

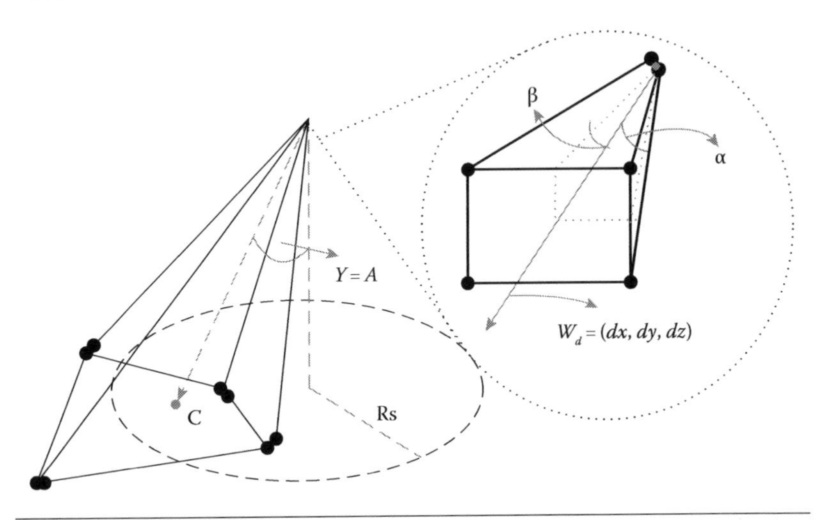

Figure 5.8 A 3D coverage model. (From Ma, H. et al., *Proceedings of the IEEE International Conference on Computer Communications*, 2009.)

in addition to the target location, the working orientation of the target has vital impact on the quality of coverage.

However, the nonexistence of commercially available coverage models was used to solve the problems of the target working orientation in the past few years, until Wang and Guohong [40] put forward the full-view coverage model. In this model, a target is recognized as full-view coverage if no matter which direction the target faces, the target is within the sensing range of at least one sensor and the target's orientation is sufficiently close to that sensor's orientation. The angle θ between the front direction of object and the sensor's orientation is a predefined parameter, we call it the effective angle. Figure 5.9 shows full-view coverage models for omnidirectional sensors and directional sensors.

5.3.2.3 Nonoverlapping/Overlapping Coverage Model In this subsection, we further divide coverage types into the overlapping model and nonoverlapping model [46], as shown in Figure 5.10. Directional sensors with the nonoverlapping coverage model only take a limited series of directions and mutually disjoint coverage sectors (see Figure 5.10a).

Compared to the nonoverlapping model [47], directional sensors with the overlapping coverage model (see Figure 5.10b) have a

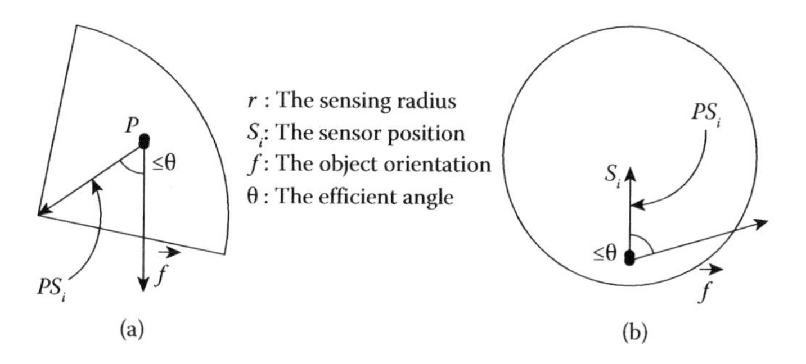

Figure 5.9 The full-view coverage models for (a) directional sensors and (b) omnidirectional sensors.

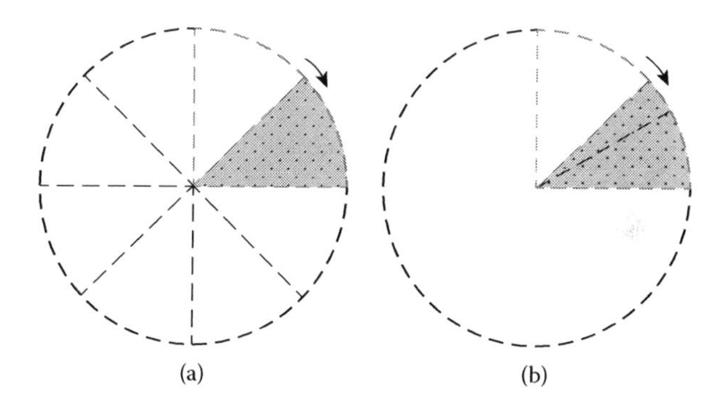

Figure 5.10 Two typical directional sensing models. (a) Nonoverlapping model and (b) Overlapping model.

limitless number of available coverage sectors based on different orientations. However, each directional node can only choose one active coverage direction at any moment. Generally speaking, the overlapping coverage model is developed to sweep all directions continuously along its center. The aim is to combine a finite set of coverage sectors to generate a whole circular viewpoint.

5.4 Factors Influencing Coverage

There are several problems (e.g., sensor deployment, movement, characteristics, location, organization) to be solved in WMSNs, because these problems generate a significant effect on coverage optimization. In this section, we mainly introduce sensor characteristics and sensor behaviors, which are the most affecting factors

Table 5.5 Research Publications on Affecting Factors in Wireless Multimedia Sensor Network Coverage

PUBLICATION	COVERAGE MODEL	SENSOR BEHAVIOR	DEPLOYMENT	COVERAGE TYPE
Wang and Cao [8]	Directional model	Static sensor	Random deployment	Barrier coverage
Yang and Qiao [39]	Omnidirectional model	Active Static sensor	Random deployment	Barrier coverage
Wang and Guohong [40]	Directional model	Static sensor	Random deployment	Area coverage
Zhang et al. [47]	Directional model	Static sensor	Deterministic deployment	Barrier coverage
Liu et al. [63]	Directional model	Static sensor	Random deployment	Target coverage
Chang et al. [64]	Directional model	Static sensor	Random deployment	Area coverage

for coverage problems. Moreover, deployment strategy, location information, and application requirements have some influence on coverage schemes. To simplify comparison, Table 5.5 lists some notable and leading publications for several different factors affecting WMSN coverage.

5.4.1 Sensor Characteristics

Compared to traditional sensors, multimedia sensors possess some unique characteristics, such as sensing radius, communication ranges, line of sight, AoV, and working direction, which bring new challenges for coverage optimization. In the following, we will introduce some of these characteristics. We have already noted these properties in the Section 5.3 except for line of sight and communication ranges; we will focus on these two characteristics in the following.

5.4.1.1 Line of Sight Unlike traditional omnidirectional sensors, each multimedia node has its own working orientation. When multimedia nodes are arranged, some issues associated with the direction need to be considered. For example, some obstacles (e.g., buildings or mountains) which are reflected in the deployment environment will affect node coverage [48]. This condition is known as the *occlusion effect*, and it directly affects the line of sight and area size for sensor coverage. The line of sight of multimedia sensors is undoubtedly decided by the size and distance between obstacles. If the intersection angle between the node and the object is 0°, we can obtain useful information from the coverage area [10]. However, the above theory is impracticable. As a result, traditional coverage models are insufficient for multimedia sensor networks.

5.4.1.2 Communication Ranges Communication range defines the farthest distance which data between nodes can be exchanged. To maximize coverage area, the basic principal is to avoid overlapping coverage where two or more multimedia sensors intersect. Existing solutions for WMSNs guarantee that communication range between nodes is at least twice that of the sensing scope [49].

5.4.2 Sensor Behaviors

In WMSNs, some sensor behaviors can often be used for achieving high coverage rates. In Subsections 5.4.2.1 through 5.4.2.3, we will introduce several characteristics of multimedia sensors: node organization, motility, and mobility.

5.4.2.1 Node Organization Generally speaking, node organization is either homogeneous or heterogeneous. A homogeneous architecture expresses that every multimedia sensor has the same processing capability. Heterogeneity is another kind of node organization that has been proposed [50]. A heterogeneous architecture expresses that some multimedia sensors are more powerful than the others. Usually, the powerful nodes are regarded as cluster heads, which can obtain information from the less powerful nodes [51]. In heterogeneous architecture, there exist multimedia sensors with different functions to meet application requirements. As a result, low-level and high-level sensors are deployed at different times or places [52].

5.4.2.2 Motility Motility—as one of the sensor behaviors—consists of actuation, pan, tilt, and zoom. Actuation yields a significant improvement in the field of sensor coverage, which is determined to actuate by cooperative relationship between neighboring multimedia sensor nodes. In Ref. [53], an intelligent sensor actuation mechanism is presented to reduce the redundancy of node information through actuating a certain number of multimedia nodes while still providing the necessary target coverage.

Pan, tilt, and zoom (PTZ), defined as motions for multimedia sensor nodes, can move and rotate the multimedia sensor along the x, y, or z axis. A PTZ multimedia sensor can randomly deploy fewer sensors at the initial phase. Meanwhile, we move and adjust multimedia sensors to meet coverage requirements. Owning to its low-cost overhead, there are a larger numbers of studies on the motility problem [54].

5.4.2.3 Mobility Motility can improve coverage performance, within reasonable delay constraints on behavior [54]. To remedy motility hole in the coverage issue, some deployment mechanisms take full advantage of mobility to redeploy nodes to sparsely covered fields after an initial random deployment [55]. Hence, mobility is feasible for improving coverage and prolonging lifetime.

In the above subchapter, we introduce the two behaviors of multimedia nodes: motility and mobility. They can improve network coverage to minimize the overlapped areas and the occlusion effect. Nevertheless, motile and mobile multimedia nodes are expensive. Consequently, Ref. [13] proposes balancing the coverage ratio and cost with hybrid multimedia sensors.

5.4.3 Other Factors

5.4.3.1 Deployment Strategy Deployment strategy plays a vital role in constructing WMSNs. Generally speaking, we construct WMSNs via two styles, deterministic deployment and random deployment. In deterministic deployment, the location and orientation of each multimedia sensor node is placed in advance. Usually, deterministic deployment can be used for a small to medium WMSN. In military applications, inhospitable surroundings, disaster sites, and remote locations, the positions and direction of sensor nodes cannot be placed in advance. In these conditions, random placement might be a good choice. Recently, a popular form of random deployment is to scatter sensors from an aircraft or to launch sensors via artillery. There has been some research in the field of random deployment [7]. Concrete deployment strategy will be explained in Section 5.5.

5.4.3.2 Location Information In WMSNs, many coverage applications including object surveillance and monitoring are dependent on location information. Hence, it is crucial to obtain location information in terms of objects and sensors. Earlier location information can be obtained through GPS (global positioning system). If it lacks the information of sensor orientation, cost, and energy, we can not obtain location data [56].

Currently, it is common to use localization algorithms to decide object and sensor location information. Sayed et al. [57] proposed a centralized algorithm. The work in Ref. [58] is an extension of Ref. [56]. Lee and Aghajan [59] presented distributed localization methods. The study in Ref. [60] obtained the position of the cameras by moving target. A localization algorithm for WMSNs was investigated in Ref. [61], which proposed an algorithm to estimate node locations to identify the areas where cameras overlap. The work presented in Ref. [62] uses 3D node localization to obtain node location information. Liu et al. [63] present a novel localization-oriented sensing model for randomly deployed nodes.

5.4.3.3 Application Requirements Application requirements for the coverage problem mainly refers to factors including coverage type, degree, ratio, and so on. Coverage type refers to the subject to be covered by nodes; coverage types can be divided into area coverage, point (target) coverage, and barrier coverage. Coverage degree describes how a target is covered. Chang et al. [64] propose k-barrier coverage. Coverage ratio measures how many nodes are needed to finish coverage in an area or how many targets satisfy the application requirement for coverage degree. Wang and Guohong [40] estimate node density for achieving full-view coverage with any given probability (e.g., 0.99). Other available studies on application requirements will be discussed in detail in Sections 5.7 and 5.8.

5.5 Sensor Deployment and Scheduling

As described in Section 5.4.3.1, sensor deployment in WMSNs can be divided into two categories, deterministic (for short deterministic deployment) and random (for short random deployment), as shown in Figure 5.11. In this section, we discuss algorithms for

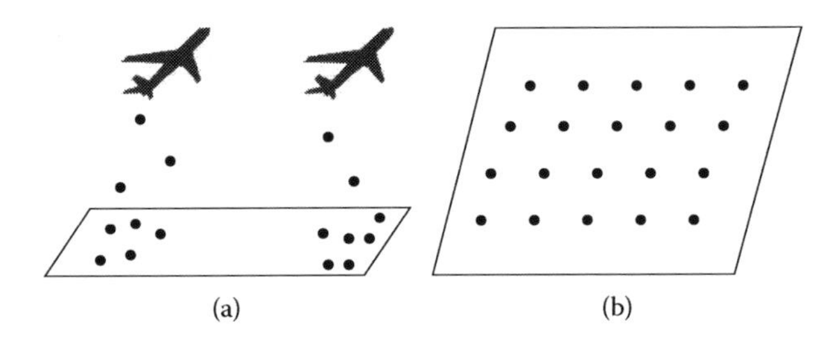

Figure 5.11 Comparison of sensor deployment. (a) Random deployment and (b) Deterministic deployment.

deterministic deployment (see Figure 5.11a), which convey that optimal node placement brings about coverage problems. Further, we present algorithms for node management and node localization in random deployment (see Figure 5.11b). Finally, we introduce several special sensor scheduling algorithms, especially for the art gallery problem.

5.5.1 Deterministic Deployment

As shown in Figure 5.11b, with deterministic deployment, MWSN nodes are deployed in an orderly fashion in predetermined locations. Through this method, we can use a minimal number of nodes to achieve maximum coverage and further reduce the cost of WMSNs [65]. However, deterministic deployment can only apply to indoor environments without any obstacles. In military applications, inhospitable or remote locations, or disaster sites, this deployment strategy is not feasible because of some overlapping and occlusion problems.

We may divide deterministic deployment into two groups: static and dynamic deployment. In static deployment, multimedia nodes can stay in the same place and cannot change their position once they are deployed [66]. The algorithm for static deployment is executed only once. If multimedia sensors can change their working directions or position after initial deployment, it is known as dynamic deployment. The algorithm in Ref. [67] provides maximal coverage through node movement with location information. In dynamic deployment, coverage can be recalculated over time, and multimedia sensors allow a new configuration to overcome the problem of coverage holes. For some scenarios, multimedia sensors can dynamically configure themselves. Ref. [68] proposes the concept of self-calibration for traditional networks. In general, a dynamic multimedia node is much more complicated than a static sensor node. Fortunately, dynamic sensors can prolong the lifetime and increase the connectivity of nodes, while maintaining or improving coverage [69].

In the past few years, research on deterministic deployment has mainly underscored working direction. A multimedia node may be deployed in *N* different directions. We may determine the best working direction for this node, in which overlapped and occlusion regions are minimized. In Ref. [70], optimal node deployment with 360° rotation is investigated. Other works that have investigated deterministic deployment have focused on camera placement (such as the art gallery problem) [71], the floodlight illumination problem [72], and the next-best view problem [73]. Recently, many studies have further discussed deterministic deployment based on more realistic assumptions. Clearly, occlusion and overlapping have significant influence on deployment problems. Mittal and Davis [74] present deterministic deployment of cameras with unchangeable orientation in a dynamic and occluded scene, which aims to cover a field of interest with minimal nodes. Similarly, Refs. [75,76] take into account the impact of the obstacle and occlusion.

The above-mentioned research has made contributions to solutions for the real-time deployment problems in covering an area of interest. In addition to deterministic deployment, there exist some additional influencing factors such as deployment structure and model. As the example of deployment model, Horster and Lienhart [77] modeled the monitored field as a grid. Zhao et al. [78] proposed a visibility model to solve the deployment problem through a binary integer programming approach. Ram et al. [79] proposed a realist node model in 3D environments. Adriaens et al. [80] found the worst-case coverage problem with camera networks. For deployment structure, the authors of Ref. [81] considered two-level deployment. Schwager et al. [82] proposed a decentralized control strategy with heterogeneous degrees of mobility. Its purpose is to use a flying robot to rotate the cameras in order to cover the monitored regions with minimal sensors.

5.5.2 Random Deployment

In military applications, inhospitable or remote locations, and disaster sites, random deployment may be the best option. Sensor nodes could be massively scattered from an aircraft or launched via artillery (see Figure 5.11b). Compared to deterministic placement, random deployment is more suitable for WMSNs. Therefore, random deployment can generate numerous redundant nodes and cause overlapped and occluded areas, while compensating for a hole in node coverage and improving fault tolerance. An interesting approach is to solve the problem using strategies such as redeployment and mobile nodes [83].

During random deployment, sensor nodes are deployed anywhere. To obtain their current location and direction, the traditional positioning system depends on GPS. The current method for node localization requires specific algorithms. There are two kinds of common localization algorithms, centralized algorithms and distributed algorithms. In centralized algorithms, the sink or a central server is used for processing data, while the other sensing node is used for capturing information. It is a useful method for saving energy. In distributed algorithms, each node needs to run all functions independently [84].

In addition, to maximize area coverage and reduce their overlap, sensor nodes adjust their current positions after initial deployment. Cai et al. [85] discuss the random deployment of nodes with changeable directions. Tezcan and Wang [86] present a distributed algorithm to adjust node direction for minimizing the effects of occlusion. Kandoth and Chellappan [87] investigate mobile multimedia sensors, where these nodes can avoid undesired overlapping and blanket spaces.

5.5.3 Sensor Scheduling

Scheduling problems have a strong correlation with coverage optimization in WMSNs. Recently, there have been several examples of sensor scheduling issues, such as the art gallery problem, robotic systems coverage, the circle covering problem, the floodlight illumination problem, the next-best view problem, and so on. In this subsection, we focus on describing the art gallery problem.

The art gallery problem is related to the concept of coverage [51]. O'Rourke [88] first gave a more formal definition for the art gallery problem. In the art gallery problem, the manager places sensors in a gallery such that the whole gallery is covered. The specific operation is to model the gallery into a polygon, such as nonoverlapping triangles [55]. This method is not feasible for a 3D space [51]; it is only suitable for a 2D space. In this model, we suppose the view angle of the node deployed in the art gallery is 360° and the sensing range is unlimited. However, this scenario is not feasible for the real world.

5.6 Sensor Management

In WMSNs, sensor nodes can obtain continuous information from the covered area with a desired quality. However, these sensor nodes can be changed as time goes by. In this scenario, the instructions and duration for sensor nodes' activity rely on the strategy of sensor management, including sensor selection and scheduling. Coverage metrics are used for evaluating the strategy of sensor management, for example, network connectivity, energy efficiency, coverage degree, coverage ratio, and so on.

A similar application on coverage metrics for selecting sensors is investigated in Ref. [89]. Similarly, Dagher et al. [90] discuss the application for complete real-time coverage of an area and provide a strategy for assigning each node's coverage area to keep minimizing energy consumption. However, this strategy is only suitable for 2D spaces without occlusions, which cannot be extended to a 3D space. Ercan et al. [91] further consider the occlusion problem.

There is currently some relevant research on sensor selection. Ref. [92] considers sensor selection with the constraints of energy costs. Considering the next-best view problem, Park et al. [93] propose a realistic 3D coverage model for sensor selection. In addition, Shen et al. [94] assign a general coverage metric to coverage space and allow task-specific weighting of the individual factors. Akyildiz and Vuran [95] present sensor selection problems. Kulkarni et al. [96] define the topological coverage overlap model with purely empirical.

The purpose of sensor scheduling is to decide the status and duration of sensor nodes, so we can guarantee the coverage requirement and prolong the network lifetime. If an area is covered by multiple sensors simultaneously, the other sensors can be considered redundant, except for a few sensors with satisfying coverage conditions. The redundant sensors should be temporarily transitioned into their energy-saving sleep state [2]. There are two node scheduling algorithms that are proposed in the literature, distributed algorithms and centralized algorithms. Because these two algorithms have been mentioned in the Section 5.5, we will not discuss them again.

5.7 Coverage Metrics

After deployment, multimedia nodes are distributed in the monitored area, with unpredicted overlapping and occlusions. Some corresponding algorithms can be used to improve the coverage of deployed nodes, but coverage benchmarks are subject to multiple factors, such as network connectivity, energy efficiency, and network lifetime. In addition, there are some other influencing factors which contain coverage performance, coverage degree, coverage ratio, and so on. In the following, we will mainly introduce several remarkable and leading coverage metrics. The related literature for coverage metrics of multimedia sensor nodes are listed in Table 5.6 to simplify comparison.

5.7.1 Network Connectivity

The coverage area of only one sensor is limited, so wireless collaboration between sensors is vital for covering larger areas [84]. Hence, network connectivity is closely related to coverage problems. It can ensure that there is at least one communication path between any two nodes. Recently, connected coverage problems have drawn much attention. Ref. [97] introduces how to determine optimal deployment to accomplish connectivity and full coverage.

The sensing model and communication radius of a node directly affects network connectivity. In WMSNs, the communication mode between nodes is omnidirectional, and most existing research assumes that the communication radius of a node is at least twice the sensing radius of this node. Under connectivity constraints, we reconsider coverage problems while assuring network connectivity. Han et al. [98] first surveyed the connected coverage problem in MWSNs; they used minimum number of directional sensors to maximize the entire target area. In addition, Ma and Liu [99] presented a deployment mechanism that maintains node coverage and network connectivity. Further, the latest research [100] solves the relationship between coverage and network connectivity in deterministic deployment.

Table 5.6 Research Publications on Coverage Metrics in Wireless Multimedia Sensor Networks

	COVERAGE METRICS				
PUBLICATION	NETWORK CONNECTIVITY	ENERGY EFFICIENCY	NETWORK LIFETIME	COVERAGE RATIO	COVERAGE DEGREE (*k*)
Kranakis et al. [97]	√	×	×	×	×
Han et al. [98]	√	×	×	√	×
Bai et al. [100]	√	×	×	×	×
Margi et al. [101]	×	√	×	√	×
Cardei and Wu [102]	√	√	√	√	√
Ai and Abouzeid [49]	×	√	√	√	√
Osais et al. [103]	√	√	×	√	×
Pescaru et al. [104]	×	√	√	√	×
Istin et al. [105]	×	√	×	√	×
Istin et al. [106]	×	√	√	√	×
Cai et al. [85]	×	√	√	×	×
Fusco and Gupta [109]	×	√	×	√	√
Kumar et al. [110]	√	×	×	√	√
Wan and Yi [111]	×	×	×	×	√
Wang et al. [112]	√	√	×	√	√
Liu et al. [113]	√	×	×	√	√
Zhao and Zheng [114]	×	×	×	√	×
Bay et al. [115]	×	×	×	√	√
Tezcan and Wang [116]	√	×	×	√	×
Chang et al. [64]	×	×	×	×	√

Note: √ indicates that coverage metrics are satisfied; × indicates that coverage metrics are not satisfied.

5.7.2 Network Lifetime

Most sensor nodes with limited battery capacity have a large influence on network lifetime. It is not feasible to replace a sensor node or recharge a battery after sensors are deployed. To prolong the network lifetime, the primary method is avoiding unnecessary energy consumption, which reflects on hardware, local processing, communication and sensing functions, and so on.

An energy-aware strategy for minimizing the energy consumption has been proposed. One pattern is to schedule redundant sensors into sleep mode. The other pattern is to adjust the transmission distance between nodes. In addition, Ref. [51] conserves energy by enhancing the efficiency of information gathering and routing. In WMSNs, scheduling redundant sensors into sleep mode or reducing the amount of active nodes is a relatively better way which can prolong network lifetime. Conversion between them may require considerable energy and time [101].

In deployment networks, especially for random deployment networks, there are a large number of redundant nodes. On one hand, redundant nodes can be used for remedying holes to improve area coverage. On the other hand, we turn off redundant nodes to save energy. The authors of Ref. [102] propose four fundamental questions for node redundancy. The first question is which sensor to put into sleep mode. The next problem is when this node should enter sleep mode; another issue is how long the node can last in this status. The last problem should consider mode feature. In addition, Ai and Abouzeid [49] present centralized and distributed algorithms to solve redundant nodes. Osais et al. [103] present an ILP model to reduce the number of sensors.

To balance coverage and network lifetime, Ref. [104] proposes the "Sensing Neighborhood Cooperative Sleeping" protocol, where the less significant nodes are shut down. Ref. [105] considers nodes with rechargeable batteries. Similar research [106] considers coverage and network lifetime with moving obstacles. Ref. [107] considers an energy-efficiency strategy based on sleeping sensor network. In Ref. [108], power management policies are defined to reduce energy consumption in MWSNs.

5.7.3 Other Coverage Metrics

5.7.3.1 Coverage Degree Depending on the application requirements, coverage degree describes how many sensors can cover an object area. The related literature proposed k-coverage, which indicates that an area is covered by at least k different nodes. For example, if a deployment field is 5-coverage, each area is covered by at least five sensor nodes, and it can accept four failed nodes while maintaining region coverage. The purpose of k-coverage is to improve coverage robustness and reliability.

There are some research papers on k-coverage problems. Ref. [109] uses a minimal number of sensors to finish k-coverage. Ref. [110]

uses k-barrier coverage to detect the target. The authors of Ref. [111] survey a k-covered region. Wang et al. [112] discuss a k-connected network problem. Liu et al. [113] use k-coverage to measure coverage performance.

5.7.3.2 Coverage Ratio As one of the coverage performance metrics, the coverage ratio measures how much area or how many targets satisfy the coverage requirement [2]. The study in Ref. [114] increased the coverage ratio by 3%. The paper [115] achieved 90% coverage. In Ref. [116], the authors studied self-orientation of multimedia nodes for maximizing coverage with occlusions. Simulations showed that the coverage ratio with occlusion-free viewpoints was significantly increased by 41%. In addition, quality of coverage, scalability, robustness, latency, jitter, adaptively, distortion, and energy consumption are also coverage metrics in WMSNs. We will not explain the coverage metrics one by one.

5.8 Typical Coverage

Cardei et al. [4] categorized available coverage problems in WMSNs into three main categories: point (target) coverage, area coverage, and barrier coverage. Point (target) coverage refers to some assigned point (the target) with a known location (see Figure 5.12). Area coverage refers to covering (monitoring) a region of interest (see Figure 5.13). Barrier coverage constructs a barrier for intrusion detection, to avoid undetected penetration, or searches for a penetration path across the monitored area (see Figure 5.14).

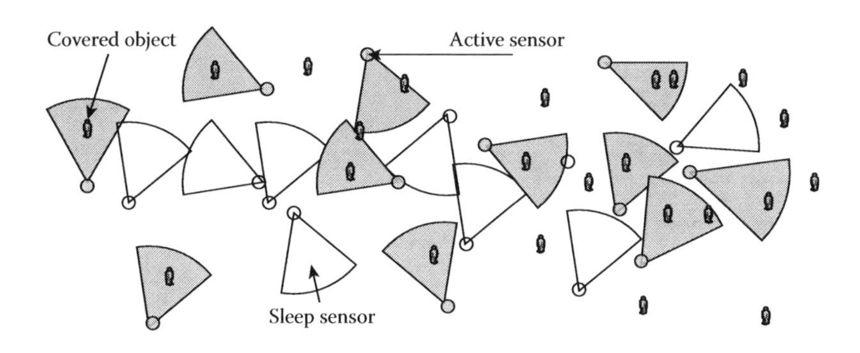

Figure 5.12 Point (target) coverage.

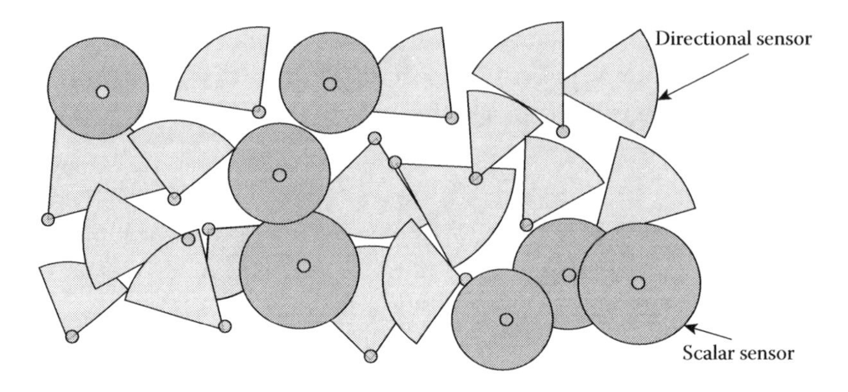

Figure 5.13 Area coverage.

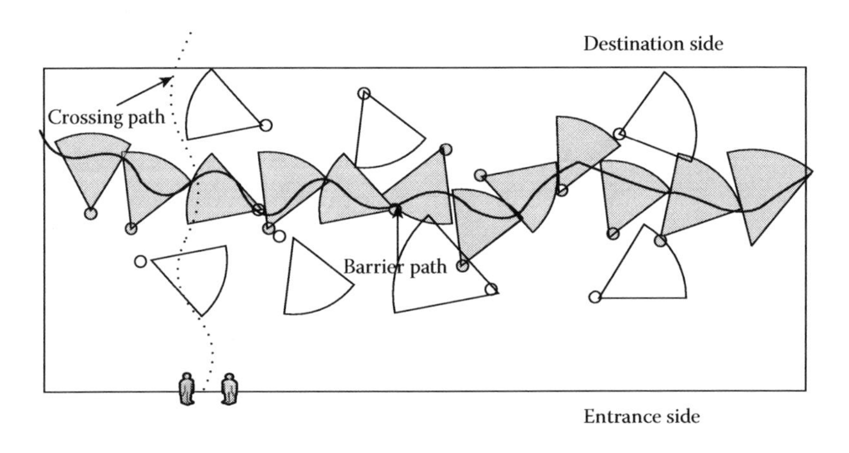

Figure 5.14 Barrier coverage.

5.8.1 Target Coverage

To cover an assigned target, researchers have defined target coverage. As shown in Figure 5.12, target coverage mainly studies how to deploy and schedule sensor nodes to satisfy maximum coverage for objects. In target coverage, each target within the monitored area is covered continuously by at least one node. Because the target within a sensor's sensing region is a discrete distribution, target coverage is also called point coverage.

There are several typical target coverage algorithms. The investigators in Ref. [117] have done extensive research to detect, classify, and track objects. The authors of Refs. [118–122] considered whether energy has an influence on target coverage. In the pioneer study of directional network coverage, Ai and Abouzeid [49] presented the problem of maximum coverage with minimum sensors problem. Considering that some deployed nodes fail and some new objects may be added as the mission requirement changes, Ref. [123] suggests reconfiguring the network by letting existing sensors steer and serve the targets periodically and solve service delay problems. To our knowledge, it is the first research on the minimum service delay problem in WMSNs.

5.8.2 Area Coverage

As shown in Figure 5.13, area coverage requires that the whole monitored area be covered by more than one node. To provide high coverage quality in WMSNs, Ref. [124] introduces the approach of utilizing a mixture of static and mobile sensors, where mobile sensors can move to make up coverage holes. To maximize multimedia sensor coverage with less time, movement distance, and message complexity, the authors of Ref. [125] use Voronoi diagrams to detect coverage holes and obtain movement trajectory. Meanwhile, Wang et al. [126] present a framework for redeployed mobile sensors in a balanced, efficient, and timely manner.

Considering directional nodes from different viewpoints can catch different views of the same objects. Ref. [40] proposes a novel concept called full-view coverage. A target point is considered to have full-view coverage if, no matter which orientation the object faces, the target is within the sensor's range and the target's facing orientation is sufficiently close to that sensor's viewing orientation. Based on the concept of full-view coverage, Ref. [52] discusses two random deployment schemes.

5.8.3 Barrier Coverage

As shown in Figure 5.14, the purpose of barrier coverage is to detect intruders that attempt to pass the monitored area. Currently, it has a large range of applications, for example: monitoring coastal waters, country borders, and boundaries of battlefields. Unlike area coverage, barrier coverage is concerned with the detectability of intruders traversing the region of interest. Unlike full coverage, barrier coverage needs a much lower number of sensors. Hence, barrier coverage in MWSNs has gained considerable attention in practice.

There have been some studies on barrier coverage, while the majority focus on traditional networks which coverage range is a circular. In recent years, there have been some requirements in WMSNs [127–129]. Compared to WSNs, the barrier coverage range for WMSNs is fan-shaped. Hence, it can capture different viewpoints for the same target point.

Barrier coverage in WSNs was first proposed in Ref. [129]. In barrier coverage, any intruding object can be identified once it passes the barrier path, regardless of the direction the object is facing. In addition, Ref. [130] proposes the maximum breach and minimum exposure path problem. In this article, the authors quantify coverage improvement when other nodes are added to the network.

Barrier coverage can be further divided into weak and strong coverage [131]. The concepts are as follows. If intruders choose the shortest path to pass through the monitored area, they can be detected. This is a weak barrier coverage. If intruders take any path from one side to the other, they can be detected; no objects can pass the undetected area. This is a strong barrier coverage.

To handle the special application requirements, Ref. [132] introduces the novel concept of full-view coverage, which we mentioned in Section 5.8.2. Based on this concept, Wang and Cao [133] construct a full-view barrier, where the monitored area is first separated into small subparts. Next, every subpart is marked up to form a graph. Further, each subpart is verified to be full-view covered. Finally, full-view covered subparts are chosen as the construct barriers.

The above distribution is not rigorous. To further reduce the number of sensors, Ma et al. [134] improved on the algorithm in Ref. [133] and proposed the minimum camera barrier coverage problem in WMSNs.

5.9 Other Research Issues

In the above-mentioned content, we have introduced some aspects of coverage problems in WMSNs. However, other currently relevant research issues are still uninvestigated. In the following content, we suggest some open research topics, such as realistic environments, 3D networking, privacy and security, node state transformation, and so on.

- Most studies of coverage problems make unrealistic assumptions (e.g., sensor model). In random deployment, the conditions in inhospitable or remote regions are not considered. Additionally, most simulation and analysis results are obtained in a fairly idealistic environment, which causes some unrealistic analysis of coverage problems. Hence, it is unreasonable for investigators to ignore these issues. Some real-world research should be considered in the future.
- To simplify coverage models, most existing research concentrates on a 2D plane. However, such network models are miles away from realistic environments. Under present conditions, research on 3D network models is still a new topic, due to high complexity in the design and analysis. These network models are largely unexplored, and the 3D network model is an extremely complex problem. As a result, the road leading to a 3D network model in WMSNs is still far away.
- WMSNs cause many challenges, while bringing gains such as people's private information. With network coverage enlarging, more and more multimedia sensors are being deployed into people's everyday lives. This circumstance causes many privacy and security issues. Certainly, there are several ways to solve this issue, such as excluding camera coverage from the network or sacrificing a bit of privacy. Nonetheless, it is still a marginal topic and is not treated seriously.
- In the area being monitored, the states of sensor nodes need to be adjusted for some application requirements. As states transformation of sensor nodes need to last for a long time, other coverage properties may be changed during this period. For example, when a multimedia sensor is switched from on to off, some routings will be broken. Coverage under such conditions is very vulnerable, and some complex and robust solutions are required to solve this issue.

5.10 Conclusions

In this article, we gave a broad survey of the ongoing research that has been solving coverage issues in WMSNs. Coverage problems can be classified into different research fields. First, we introduced and classified existing available WMSN hardware. Moreover, we presented several coverage models, which can measure sensing capability and quality through the geometric relation between objects and sensor nodes.

Second, we further divided coverage problems into four stages: the design stage, deployment stage, management stage, and coverage metrics. In the design stage, we noted some factors that can significantly influence coverage optimization. In the deployment stage, we mainly considered deterministic and random deployment. In the management stage, we discussed scheduling different sensor nodes to work alternatively to prolong network lifetime while preserving network coverage. In addition, we also related benchmarks for measuring coverage performance, such as network connectivity, energy consumption, and so on.

Finally, we touched on several typical coverage classifications including area coverage, target coverage, and barrier coverage. We reviewed representative solution algorithms in each category. In addition, coverage problems are still a recent, open research topic. We surveyed and classified existing recent studies on topics such as realistic environments, 3D networks, privacy and security, and node state transformation for WMSNs. We hope our work is useful for other researchers.

References

1. Akyildiz, I.F., Melodia, T., Chowdhury, K.R. Wireless multimedia sensor networks: Applications and testbeds. *Proceedings of the IEEE* 96(10), 1588–1605, 2008.
2. Wang, B. Coverage problems in sensor networks: A survey. *ACM Computing Surveys* 43(4), 32, 2011.
3. Liu, H., Wan, P., Yi, C.W., Jia, X., Makki, S., Pissinou, N. Maximal lifetime scheduling in sensor surveillance networks. *The 24th Annual Joint Conference of the IEEE Computer and Communications Societies*, pp. 2482–2491, 2005. DOI: 10.1109/INFCOM.2005.1498533.
4. Cardei, M., Thai, M., Li, Y., Wu, W. Energy-efficient target coverage in wireless sensor networks. *The 24th Annual Joint Conference of the IEEE Computer and Communications Societies*, pp. 1976–1984, 2005. DOI:10.1109/INFCOM.2005.1498475.

5. Huang, C.F., Tseng, Y.C. A survey of solutions to the coverage problems in wireless sensor networks. *Journal of Internet Technology* 6, 1–8, 2005.
6. Megerian, S., Koushanfar, F., Potkonjak, M., Srivastava, M. Worst and best-case coverage in sensor networks. *IEEE Transactions on Mobile Computing* 4, 1, 84–92, 2005. DOI: 10.1109/TMC.2005.1(410).
7. Ma, H., Liu, Y. On coverage problems of directional sensor networks. In: Lecture Notes in Computer Science. *Mobile Ad-Hoc and Sensor Networks* 3794, 721–731, 2005.
8. Wang, Y., Cao, G. Achieving full-view coverage in camera sensor networks. *ACM Transactions on Sensor Networks*, 2013.
9. Soro, S., Heinzelman, W. On the coverage problem in video-based wireless sensor networks. In: *Proceedings of the 2nd International Conference on Broadband Networks*, pp. 932–939, IEEE: Boston, MA, USA, October 3–7, 2005.
10. Akyildiz, I.F., Melodia, T., Chowdhury, K.R. A survey on wireless multimedia sensor networks. *Computer Networks* 51, 4, 921–960, 2007.
11. Soro, S., Heinzelman, W. A survey of visual sensor networks. *Hindawi Advances in Multimedia* 2009. DOI: 10.1155/2009/640386.
12. Greenleaf, A., Photographic Optics, Macmillan, London, England, 1950.
13. Amac Guvensan, M., Gokhan Yavuz, A. On coverage issues in directional sensor networks: A survey. *Ad Hoc Networks* 9, 1238–1255, 2011.
14. Carnegie Mellon University, CMUcam3 Datasheet Version 1.02, Pennsylvania, September 2007.
15. Yap, F.G.H., Yen, H.H. A survey on sensor coverage and visual data capturing/processing/transmission in wireless visual sensor networks. *Sensors* 14, 3506–3527, 2014. DOI: 10.3390/s140203506.
16. Omnivision Technologies. http://www.ovt.com
17. Agilent Technologies. http://www.home.agilent.com
18. Wise GEEK. http://www.wisegeek.org/what-is-an-infrared-sensor.htm
19. Wise GEEK. http://www.wisegeek.com/what-is-a-motion-detector.htm
20. eHow. http://www.ehow.com/how-does_5561845_do-ir-sensors-work.html
21. azosensors. http://www.azosensors.com/Article.aspx?ArticleID=339
22. Wise GEEK. http://www.wisegeek.com/what-is-an-ultrasonic-sensor.htm
23. eHow. http://www.ehow.com/how-does_4947693_ultrasonic-sensors-work.html
24. Almalkawi, I.T., Guerrero Zapata, M., Al-Karaki, J.N., Morillo-Pozo, J. Wireless multimedia sensor networks: current trends and future directions. *Sensors* 10, 6662–6717, 2010. DOI: 10.3390/s100706662.
25. Rahimi, M., Baer, R., Iroezi, O., Garcia, J., Warrior, J., Estrin, D., Srivastava, M. Cyclops: In situ image sensing and interpretation in wireless sensor networks. In: *Proceedings of the 3rd ACM Conference Embedded Network Sensor Systems (SenSys)*, ACM Press, San Diego, CA, USA, Nov. 2005.
26. Carnegie Mellon University. CMUcam3 datasheet, version 1.02. Pittsburgh, PA, USA, Sep. 2007.

27. The Stargate Platform. http://www.xbow.com/Products/Xscale.htm
28. Teixeira, T., Culurciello, E., Park, J., Lymberopoulos, D., Barton-Sweeney, A., Savvides, A. Address-event imagers for sensor networks: Evaluation and modeling. In: *Proceedings of the 5th International Conference on Information Processing in Sensor Networks, IPSN 2006*, pp. 458–466, IEEE: Nashville, TN, USA, April, 2006.
29. Kulkarni, P., Ganesan, D., Shenoy, P., Lu, Q. SensEye: A multi-tier camera sensor network. In: *Proceedings of the 13th Annual ACM International Conference on Multimedia, MULTIMEDIA'05*. pp. 229–238, ACM: New York, NY, USA, 2005.
30. Feng, W.C., Kaiser, E., Feng, W.C., Baillif, M.L. Panoptes: Scalable low-power video sensor networking technologies. *ACM Transactions on Multimedia Computing, Communications, and Applications* 1, 151–167, 2005.
31. Hengstler, S., Prashanth, D., Fong, S., Aghajan, H. MeshEye: A hybrid-resolution smart camera mote for applications in distributed intelligent surveillance. In: *Proceedings of the International Conference on Information Processing in Sensor Networks (IPSN)*, pp. 360–369, IEEE: Cambridge, MA, USA, 2007.
32. Kurkowski, S., Camp, T., Colagrosso, M., MANET simulation studies: The incredibles. *ACM SIGMOBILE Mobile Computing and Communications Review* 9(4), 50–61, 2005.
33. Hengstler, S., Aghajan, H. WiSNAP: A wireless image sensor network application platform. In: *Proceedings of 2nd International Conference on Testbeds and Research Infrastructures for the Development of Networks and Communities, TRIDENTCOM 2006*, pp. 6–12, Barcelona, Spain, 1–3 March 2006.
34. Boice, J., Lu, X., Margi, C., Stanek, G., Zhang, G., Obraczka, K. Meerkats: A power-aware, self-managing wireless camera network for wide area monitoring. In: *Distributed Smart Cameras Workshop - SenSys06*, Boulder, CA, USA, 2006.
35. Campbell, J., Gibbons, P.B., Nath, S., Pillai, P., Seshan, S., Sukthankar, R. IrisNet: An Internet-scale architecture for multimedia sensors. In: *Proceedings of the 13th annual ACM International Conference on Multimedia, MULTIMEDIA'05*, pp. 81–88, ACM: New York, NY, USA, 2005.
36. Tavli, B., Bicakci, K., Zilan, R., Barcelo-Ordinas, J.M. A survey of visual sensor network platforms. *Multimedia Tools and Applications* 60, 689–726, 2012.
37. Wang, Y. Coverage Problems in Camera Sensor Networks. *The Pennsylvania State University*, Pennsylvania, 2013.
38. Onur, E., Ersoy, C., Delic, H., Akarun, L. Surveillance wireless sensor networks: Deployment quality analysis. *Network, IEEE* 21, 6, 48–53. DOI: 10.1109/*MNET.2007*.4395110, 2007.
39. Yang, G., Qiao, D. Barrier information coverage with wireless sensors. In: *Proceedings of the IEEE Infocom Conference on Computer Communications*, IEEE: Rio de Janeiro, Brazil, 918–926, 2009.

40. Wang, Y., Guohong, C. On full-view coverage in camera sensor networks. *IEEE International Conference on Computer Communications* 1781–1789, 2011.
41. Huang, C.F., Tseng, Y.C. The coverage problem in a wireless sensor network. *Mobile Network and Applications* 519–528, 2005.
42. Shen, C., Cheng, W., Liao, X., Peng, S. Barrier coverage with mobile sensors. In: *International Symposium on Parallel Architectures, Algorithms, and Networks*, pp. 99–104, 2008.
43. Kloder, S., Hutchinson, S. Barrier coverage for variable bounded-range line-of-sight guards, in Robotics and Automation. *IEEE International Conference* 391–396, 2007.
44. Kloder, S., Hutchinson, S. Partial barrier coverage: Using game theory to optimize probability of undetected intrusion in polygonal environments, *2008 IEEE International Conference on Robotics and Automation, Pasadena, CA, USA*, May 19–23, 2671–2676, 2008.
45. Ma, H., Zhang, X., Ming, A. A coverage-enhancing method for 3d directional sensor networks. In: *Proceedings of the 28th IEEE International Conference on Computer Communications (INFOCOM'09)*, pp. 2791–2795, IEEE: Rio de Janerio, Brazil, 2009. DOI: 10.1109/INFCOM.2009.5062233.
46. Tao, D., Ma, H. Coverage control algorithms for directional sensor networks. *Journal of Software* 22(10), 2315–2332, 2011.
47. Zhang, L., Tang, J., Zhang, W. Strong barrier coverage with directional sensors. In: *Proceedings of IEEE Global Telecommunications Conference (Globecom)*, 1–6, IEEE: Honolulu, HI, USA, 2009. DOI: 10.1109/GLOCOM.2009.5425893.
48. Tezcan, N., Wang, W. Self-orienting wireless multimedia sensor networks for occlusion-free viewpoints. *Computer Networks: International Journal of Computer and Telecommunications Networking* 52(13), 2558–2567, 2008. DOI: http://dx.doi.org/10.1016/j.comnet.2008.05.014.
49. Ai, J., Abouzeid, A.A. Coverage by directional sensors in randomly deployed wireless sensor networks. *Journal of Combinatorial Optimization* 11(1), 21–41, 2006. DOI: 10.1007/s10878-006-5975-x.
50. Wang, X., Wang, X., Zhao, J. Impact of mobility and heterogeneity on coverage and energy consumption in wireless sensor networks. In: *Proceedings of IEEE ICDCS 2011*, pp. 477–487, Minneapolis, MN, USA, June 21–24, 2011.
51. Mulligan, R., Ammari, H.M. Coverage in wireless sensor networks: A survey. *Network Protocols and Algorithms*. ISSN 1943-3581. 2(2), 2010.
52. Wu, Y., Wang, X. Achieving full view coverage with randomly-deployed heterogeneous camera sensors. *The 32nd IEEE International Conference on Distributed Computing Systems*, pp. 556–565, 2012. DOI: 10.1109/ICDCS.2012.9.
53. Devarajan, D., Radke, R. Calibrating distributed camera networks using belief propagation. *EURASIP J. Appl. Signal Process.* 2007.

54. Kansal, A., Kaiser, W.J., Pottie, G.J., Srivastava, M.B. Actuation techniques for sensing uncertainty reduction, Technical Reports, pp. 1–16, 2005.
55. Fan, G.J., Jin, S.Y. Coverage problem in wireless sensor network: A survey. *Journal of Networks* 5(9), 2010. DOI: 10.4304/jnw.5.9.1033-1040.
56. Fuiorea, D., Guia, V., Pescaru, D., Toma, C. Using registration algorithms for wireless sensor network node localization. In: *Proceedings of 4th IEEE International Symposium on Applied Computational Intelligence and Informatics*, pp. 209–214, IEEE: Timisoara, Romania, May 17–18, 2007.
57. Sayed, A.H., Tarighat, A., Khajehnouri, N. Network-based wireless location: Challenges faced in developing techniques for accurate wireless location information. Signal Processing Magazine, IEEE 22, 24–40, 2005.
58. Fuiorea, D., Gui, V., Pescaru, D., Paraschiv, P., Codruta, I., Curiac, D., Volosencu, C. Video-based wireless sensor networks localization technique based on image registration and SIFT algorithm. *WSEAS Transactions on Computers*, pp. 990–999, 2008.
59. Lee, H., Aghajan, H. Vision-enabled node localization in wireless sensor networks. In: *Proceedings of Cognitive Systems with Interactive Sensors*, pp. 1–8. ACM Press, Paris, France, March 15–17, 2006.
60. Funiak, S., Paskin, M., Guestrin, C., Sukthankar, R. Distributed localization of networked cameras. In: *Proceedings of the 5th International Conference on Information Processing in Sensor Networks*, pp. 34–42, IEEE: Nashville, TN, USA, April 19–21, 2006.
61. Shafique, K., Hakeem, A., Javed, O., Haering, N. Self calibrating visual sensor networks. In: *Proceedings of IEEE Workshop on Applications of Computer Vision*, pp. 1–6, Copper Mountain, CO, USA, January 7–9, 2008.
62. Barton-Sweeney, A., Lymberopoulos, D., Savvides, A. Sensor localization and camera calibration in distributed camera sensor networks. In: *Proceedings of the 3rd International Conference on Broadband Communications, Networks and Systems*, pp. 1–10, IEEE: San Jose, CA, USA, October 1–5, 2006.
63. Liu, L., Zhang, X., Ma, H. Localization-oriented coverage in wireless camera sensor networks. *IEEE Transactions on Wireless Communications* 10(2), pp. 484–494, 2011.
64. Chang, C.Y., Hsiao, C.Y., Chang, C.T. The k-barrier coverage mechanism in wireless visual sensor networks. *IEEE Wireless Communications and Networking Conference: Mobile and Wireless Networks*, pp. 2318–2322, 2012.
65. Osais, Y.E., St-Hilaire, M., Riu, F.R. Directional sensor placement with optimal sensing ranging, field of view and orientation. *Mobile Networks and Applications* 15, 216–225, 2010.
66. Younis, M., Akkaya, K. Strategies and techniques for node placement in wireless sensor. networks: A survey. *Ad Hoc Networks*, pp. 621–655, 2008.

67. Howard, A., Mataric, M.J., Sukhatme, G.S. An incremental self deployment algorithm for mobile sensor networks. *Autonomous Robots* 13(2), 113–126, 2002.
68. Zhou, Z., Das, S., Gupta, H. Variable radii connected sensor cover in sensor networks. *ACM Transactions on Sensor Networks*, pp. 1–36, 2009.
69. Gasparri, A., Krishnamachari, B., Sukhatme, G.S. A framework for multi-robot node coverage in sensor networks. *Annals of Mathematics and Artificial Intelligence* 52(2–4), 281–305, 2008.
70. Couto, M., Souza, C., Rezende, P. Strategies for optimal placement of surveillance cameras in art galleries. In: *Proceedings of 18th International Conference on Computer Graphics and Vision*, pp. 1–4, Moscow, Russia, June 23–27, 2008.
71. Marengoni, M., Draper, B., Handson, A., Sitaraman, R. A system to place observers on a polyhedral terrain in a polynomial time. *Image and Vision Computing* 773–780, 1996.
72. Bose, P., Guibas, L., Lubiw, A., Overmars, M., Souvaine, D., Urrutia, J. The floodlight problem. *International Journal of Computational Geometry & Applications* 153–163, 1997.
73. Pito, R. A solution to the next best view problem for automated surface acquisition. *IEEE Transactions on Pattern Analysis and Machine Intelligence* 21, 1016–1030, 1999.
74. Mittal, A., Davis, L. Visibility analysis and sensor planning in dynamic environments. In: *Proceedings of 8th European Conference on Computer Vision*, pp. 175–189, Prague, Czech Republic, May 11–14, 2004.
75. Lin, Y.T., Saluja, K.K., Megerian, S. Adaptive cost efficient deployment strategy for homogeneous wireless camera sensors. *Ad Hoc Networks* 9, 713–726, 2011.
76. Karakaya, M., Qi, H. Coverage estimation for crowded targets in visual sensor networks. *ACM Transactions on Sensor Networks*, pp. 41–49, 2012.
77. Horster, E., Lienhart, R. Approximating optimal visual sensor placement. In: *Proceedings of IEEE International Conference on Multimedia and Expo*, pp. 1257–1260, Toronto, ON, Canada, July 9–12, 2006.
78. Zhao, J., Cheung, S., Nguyen, T. Optimal camera network configurations for visual tagging. *IEEE Journal of Selected Topics in Signal Processing* 464–479, 2008.
79. Ram, S., Ramakrishnan, K., Atrey, P., Singh, V., Kankanhalli, M. A design methodology for selection and placement of sensors in multimedia surveillance systems. In: *Proceedings of the 4th ACM International Workshop on Video Surveillance and Sensor Networks*, pp. 121–130, Santa Barbara, CA, USA, October 27, 2006.
80. Adriaens, J., Megerian, S., Pontkonjak, M. Optimal worst-case coverage of directional field-of-view sensor networks. In: *Proceedings of 3rd Annual IEEE Communications Society Conference on Sensor, Mesh and Ad Hoc Communications and Networks*, pp. 336–345, Reston, VA, USA, September 25–28, 2006.

81. Li, H., Pandit, V., Agrawal, D.P. Deployment optimization strategy for a two-tier wireless visual sensor network. *Wireless Sensor Network* 91–106, 2012.
82. Schwager, M., Julian, B., Angermann, M., Rus, D. Eyes in the sky: Decentralized control for the deployment of robotic camera networks. *Proceedings of the IEEE* 1541–1561, 2011.
83. Pescaru, D., Gui, V., Toma, C., Fuiorea, D. Analysis of post-deployment sensing coverage for video wireless sensor networks. In: *Proceedings of 6th International Conference RoEduNet*, Craiova, Romania, November 23–24, 2007.
84. Costa, D.G., Guedes, L.A. The coverage problem in video-based wireless sensor networks: A survey. *Sensors 2010*, 10, 8215–8247, 2010. DOI: 10.3390/s100908215.
85. Cai, Y., Lou, W., Li, M., Li, X.Y. Target-oriented scheduling in directional sensor networks. In: *Proceedings of IEEE Infocom*, pp. 1550–1558, Anchorage, AK, USA, May 6–12, 2007.
86. Tezcan, N., Wang, W. Self-orienting wireless sensor networks for occlusion-free viewpoints. *Computer Networks* 52, 2558–2567, 2008.
87. Kandoth, C., Chellappan, S. Angular mobility assisted coverage in directional sensor networks. In: *Proceedings of International Conference on Network-Based Information Systems*, pp. 376–379, Indianapolis, IN, USA, August 19–21, 2009.
88. O'Rourke, J. *Computational Geometry in C.* Cambridge University Press, Cambridge, Britain, March 25, 1994.
89. Soro, S., Heinzelman, W. Camera selection in visual sensor networks. In: *Proceedings of the IEEE Conference on Advanced Video and Signal Based Surveillance (AVSS '07)*, pp. 81–86, 2007.
90. Dagher, J.C., Marcellin, M.W., Neifeld, M.A. A method for coordinating the distributed transmission of imagery. *IEEE Transactions on Image Processing* 15(7), 1705–1717, 2006.
91. Ercan, A., Gamal, A.E., Guibas, L. Camera network node selection for target localization in the presence of occlusions. In: *Proceedings of the ACM SenSys Workshop on Distributed Smart Cameras*, 2006.
92. Mavrinac, A., Chen, X. Modeling coverage in camera networks: A survey. *Springer Science Business Media New York* 2012. DOI: 10.1007/s11263-012-0587-7.
93. Park, J., Bhat, P.C., Kak, A.C. A look-up table based approach for solving the camera selection problem in large camera networks. In: *Proceedings of International Workshop on Distributed Smart Cameras*, 2006.
94. Shen, C., Zhang, C., Fels, S. A multi-camera surveillance system that estimates quality-of-view measurement. In: *Proceedings of IEEE International Conference on Image Processing*, pp. 193–96, 2007.
95. Akyildiz, I.F., Vuran, M.C. *Wireless Sensor Networks*. John Wiley and Sons, 2010.
96. Kulkarni, P., Shenoy, P., Ganesan, D. Approximate initialization of camera sensor networks. In: *Proceedings of 4th European Conference on Wireless Sensor Networks*, pp. 67–82, 2007.

97. Kranakis, E., Krizanc, D., Urrutia, J. Coverage and connectivity in networks with directional sensors. In: *Euro-Par 2004 Parallel Processing, Lecture Notes in Computer Science*, vol. 3149, pp. 917–924, Springer: Berlin, Heidelberg, 2004.
98. Han, X., Cao, X., Lloyd, E., Shen, C.C. Deploying directional sensor networks with guaranteed connectivity and coverage. In: *Proceedings of 5th Annual IEEE Communications Society Conference on Sensor, Mesh and Ad Hoc Communications and Networks (SECON'08)*, pp. 153–160, San Francisco, CA, USA, 2008. DOI: 10.1109/SAHCN.2008.28.
99. Ma, H., Liu, Y. Some problems of directional sensor networks. *International Journal of Sensor Networks* 2(1/2), 44–52, 2007.
100. Bai, X., Yun, Z., Xuan, D., Lai, T.H., Jia, W. Optimal patterns for four-connectivity and full coverage in wireless sensor networks. *IEEE Transactions on Mobile Computing* 9(3), 435–448, 2010.
101. Margi, C.B., Manduchi, R., Obraczka, K. Energy consumption tradeoffs in visual sensor networks. In: *Proceedings of 24th Brazilian Symposium on Computer Networks*, Curitiba, Brazil, May 2006.
102. Cardei, M., Wu, J. Energy-efficient coverage problems in wireless ad hoc sensor networks. *Computer Communications* 413–420, 2006.
103. Osais, Y., St-Hilaire, M., Yu, F. Directional sensor placement with optimal sensing range, field of view and orientation. In: *Proceedings of IEEE International Conference on Wireless and Mobile Computing (WIMOB'08)*, pp. 19–24, Avignon, France, 2008. DOI: 10.1109/WiMob.2008.88.
104. Pescaru, D., Istin, C., Curiac, D., Doboli, A. Energy saving strategy for video-based wireless sensor networks under field coverage preservation. In: *Proceedings of IEEE International Conference on Automation, Quality and Testing, Robotics*, pp. 289–294, Cluj-Napoca, Romania, May 22–25, 2008.
105. Istin, C., Pescaru, D., Ciocarlie, H., Curiac, D., Doboli, A. Reliable field of view coverage in video-camera based wireless networks for traffic management applications. In: *Proceedings of IEEE International Symposium on Signal Processing and Information Technology*, pp. 63–68, Sarajevo, Bosnia and Herzegovina, December 16–19, 2008.
106. Istin, C., Pescaru, D., Doboli, A., Ciocarlie, H. Impact of coverage preservation techniques on prolonging the network lifetime in traffic surveillance applications. In: *Proceedings of 4th International Conference on Intelligent Computer Communication and Processing*, pp. 201–206, Cluj-Napoca, Romania, August 28–30, 2008.
107. Kumar, S., Lai, T.H., Balogh, J. On k-coverage in a mostly sleeping sensor network. *MobiCom'04*, Sept. 26–Oct. 1, 2004.
108. Misra, S., Reisslein, M., Xue, G. A survey of multimedia streaming in wireless sensor networks. *IEEE Communications Surveys & Tutorials* 10(4), 18–39, 2008.
109. Fusco, G., Gupta, H. Selection and orientation of directional sensors for coverage maximization. In: *Proc. of IEEE Intl. Conf. on Sensor, Mesh and Ad Hoc Communications and Networks (SECON'09)*, Rome, Italy, pp. 1–9, 2009. DOI: 10.1109/SAHCN.2009.5168968.

110. Kumar, S., Lai, T.H., Arora, A. Barrier coverage with wireless sensors. In: *Proc. of ACM Intl. Conf. on Mobile computing and networking (MobiCom'05)*, pp. 284–298, 2005. DOI: http://doi.acm.org/10.1145/1080829.1080859.
111. Wan, P., Yi, C. Coverage by randomly deployed wireless sensors networks. *IEEE Transactions on Information Theory* 52, 2658–2669, 2006.
112. Wang, X., Xing, G., Zhang, Y., Lu, C., Pless, R., Gill, C. Integrated coverage and connectivity configuration in wireless sensor networks. In: *Proceedings of 1st ACM Conference on Embedded Networked Sensor Systems*, pp. 28–39, Los Angeles, CA, USA, November, 2003.
113. Liu, L., Ma, H., Zhang, X. On directional k-coverage analysis of randomly deployed camera sensor networks. In: *Proceedings of IEEE International Conference on Communications*, pp. 2707–2711, Beijing, China, May 19–23, 2008.
114. Zhao, J., Zeng, J.C. An electrostatic field-based coverage-enhancing algorithm for wireless multimedia sensor networks. In: *Proc. of IEEE Intl. Conf. on Wireless Communications, Networking and Mobile Computing (WiCom'09)*, pp. 1–5, Beijing, China, 2009. DOI: 10.1109/WICOM.2009.5302443
115. Bay, H., Ess, A., Tuytelaars, T., Van Gool, L. Speeded-up robust features (surf). *Computer Vision and Image Understanding* 110(3), 2010.
116. Tezcan, N., Wang, W. Self-orienting wireless multimedia sensor networks for maximizing multimedia coverage. *IEEE Communications Society ICC*, 2206–2210, 2008.
117. Arora, A., Dutta, P. et. al. A line in the sand: A wireless sensor network for target detection, classification, and tracking, *Computer Networks. Computer and Telecommunications Networking*, pp. 605–634, 2004.
118. Cardei, M., Thai, M., Yingshu, L., Weili, W. Energy-efficient target coverage in wireless sensor networks. *INFOCOM 2005*, 1976–1984, 2005.
119. Cardei, M., Du, D. Improving wireless sensor network lifetime through power aware organization. *Wireless Networks*, pp. 330–333, 2005.
120. Cardei, M., Wu, J., Lu, M., Pervaiz, M. Maximum network lifetime in wireless sensor networks with adjustable sensing ranges. *Wireless and Mobile Computing, Networking And Communications*, pp. 438–445, 2005.
121. Zhang, H., Wang, H., Feng, H. A distributed optimum algorithm for target coverage in wireless sensor networks. *Asia-Pacific Conference on Information Processing*, pp. 144–147, 2009.
122. Zhang, H. Energy-balance heuristic distributed algorithm for target coverage in wireless sensor networks with adjustable sensing ranges. *Asia-Pacific Conference on Information Processing*, pp. 452–455, 2009.
123. Wang, Y., Cao, G. Minimizing service delay in directional sensor networks. *Computer Communications. Piscataway* 2011, 1790–1798, 2011.
124. Wang, G., Cao, G., LaPorta, T. A bidding protocol for deploying mobile sensors. *ICNP'03*, pp. 315–324, 2003.
125. Wang, G., Cao, G., LaPorta, T. Movement-assisted sensor deployment. *IEEE INFOCOM*, 2004.

126. Wang, G., Cao, G., LaPorta, T., Zhang, W. Sensor relocation in mobile sensor networks. *IEEE INFOCOM*, 2005.
127. Devarajan, D., Ranke, R.J., Chung, H. Distributed metric calibration of ad-hoc camera networks. *ACM Transactions on Sensor Networks*, pp. 33–44, 2006.
128. Johnson, M.P., Bar-Noy, A. Pan and scan: Configuring cameras for coverage. *INFOCOM*, pp. 1071–1079, 2011.
129. Shih, K.P., Chou, C.M., Liu, I.H., Li, C.C. On barrier coverage in wireless camera sensor networks. *IEEE AINA*, pp. 873–879, 2010.
130. Meguerdichian, S., Koushanfar, F., Potkonjak, M., Srivastava, M. Coverage problems in wireless ad-hoc sensor networks. *IEEE Infocom 2001*, 1380–1387, 2001.
131. Chen, A., Kumar, S., Lai, T.H. Designing localized algorithms for barrier coverage. In: *MOBICOM*, 2007.
132. Balasubramaniam, S., Kangasharju, J. Realizing the Internet of nano things: Challenges, solutions, and applications. *IEEE Computer Society* 62–68, 2013.
133. Wang, Y., Cao, G. Barrier coverage in camera sensor networks. *ACM International Symposium on Mobile Ad Hoc Networking and Computing*, May 16–19, 2011.
134. Ma, H., Yang, M., Li, D. Minimum camera barrier coverage in wireless camera sensor networks. *IEEE INFOCOM 2012*, 217–225, 2012.

6

A Security Scheme for Video Streaming in Wireless Multimedia Sensor Networks

NOUR EL-DEEN MAHMOUD KHALIFA

Contents

Wireless sensor networks (WSNs) have become an important component of our daily life. In the near future, they will dominate the technology industry around the world. WSNs have gained importance due to the variety of vital applications they can participate in, such as applications for the military, health care, agriculture, surveillance, and monitoring natural phenomena.

Wireless multimedia sensor networks (WMSNs) are a special type of WSN. WMSNs typically consist of wirelessly interconnected devices that are able to ubiquitously retrieve multimedia content such as video and audio streams, still images, and scalar sensor data from the environment. These wireless devices, called sensor nodes, are limited in energy and storage capabilities. The sensor nodes collect data from physical or environmental phenomena. They cooperatively pass the sensed data through the network to a certain location or sink node where the data can be collected and analyzed.

The unprotected nature of wireless communication channels and untrustworthy transmission medium of WMSNs are vulnerable to many types of security attacks. Attackers ultimately seek to eavesdrop, steal confidential data, inject false data, or even jam the whole network, so securing these networks has become essential.

This chapter will propose a security scheme for WMSNs. The proposed security scheme is appropriate for the nature of real-time video streaming. It constructs its security features within the application and transport layers, because the information that attackers ultimately seek exists within these layers. The proposed security scheme consists of two security levels. The first level is encryption of the packet data by means of applying the Advanced Encryption Standard (AES); http://csrc.nist.gov/publications/fips/fips197/fips-197.pdf, while the second level consists of generating a Message Authentication Code (MAC) using Cipher-based Message Authentication Code (CMAC). The rationale for using both these security techniques will be justified in detail in this chapter. Both levels achieve the principles of security—authentication, confidentiality, data integrity, and availability.

Performance comparisons between the proposed security scheme and other security frameworks will be presented. All the work presented in this chapter was developed and implemented using Network Simulator version 2 (NS2). According to our literature review, this study is one of the first attempts to use NS2 as a security simulator; NS2 has not supported security features before.

Finally, it is hoped that the research presented in this chapter will help other researchers to simulate their security ideas and mechanisms based on the suggested development and integration of security libraries into NS2.

6.1 Wireless Multimedia Sensor Networks

A WMSN is composed of low cost, low power, multifunctional sensor nodes (Figure 6.1), which are small in size and communicate wirelessly over short distances. WMSNs can also be introduced as self-configured wireless networks used to collect data (video, image, audio, and scalar sensor data) from physical or environmental phenomena and deliver the multimedia content through sensors [1]. The networks cooperatively pass their sensed data through the network to a main location or sink where the data can be monitored and analyzed. A sink node or base station performs like a gate between users and the network [1] (Figure 6.2).

Figure 6.1 An example of a wireless multimedia sensor node. (From Akyildiz, I.F., and Vuran, M.C.: *Wireless sensor networks*. 2010. Copyright Wiley-VCH Verlag GmbH & Co. KGaA.)

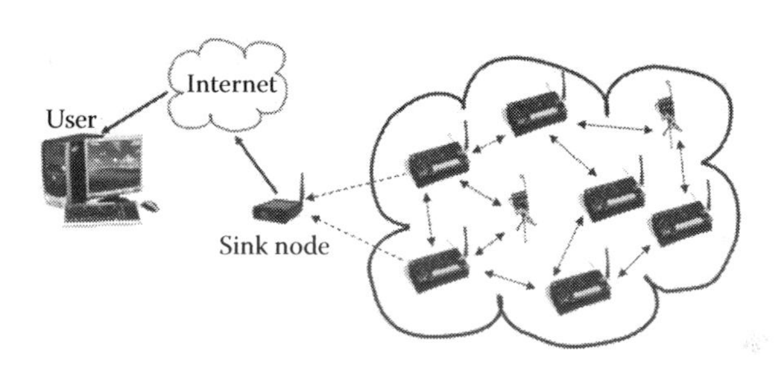

Figure 6.2 A generic wireless multimedia sensor network architecture. (From Akyildiz, I.F., and Vuran, M.C.: *Wireless sensor networks*. 2010. Copyright Wiley-VCH Verlag GmbH & Co. KGaA.)

In general, a WMSN may contain hundreds or thousands of sensor nodes. The sensor nodes can communicate among themselves using radio signals as illustrated in Figure 6.2. After the sensor nodes are deployed in the area to be monitored, they are accountable for self-organizing an acceptable network infrastructure, often using multihop communication with each other. They then start gathering information of interest [1].

There is greater flexibility in deployment options for WMSNs compared to other networks. Once WMSN nodes are distributed randomly in an environment, they organize themselves into a coherent information-sharing network. WMSN nodes may also be deployed in an organized way by creating a network topology before deploying them in the actual environment. Wireless sensor devices also respond to orders sent from an “administration site” to do specific instructions or provide sensing data samples. The working style of the sensor nodes may be either continuous or event driven [1].

6.1.1 WMSN Node Structure

A WMSN node consists of five main components: a processing unit, memory, a transceiver, sensors (video camera), and a power supply, as shown in Figure 6.3. Every unit in the WMSN node has specific tasks to be done. The following points will illustrate those tasks [2]:

- The *transceiver unit* allows the node to communicate with neighboring nodes.
- The *sensor unit* consists of two parts. First, there is an analog sensing component such as a video camera, which physically measures environmental characteristics. Second, there is an analog-to-digital converter, which transforms analog environmental readings into a digital representation that can be handled by the node processor.
- The *power source unit* provides the node with life and is normally limited, so that once the node’s power supply is exhausted the node can no longer operate.

A node may contain other elements, such as power-generating components (e.g., solar panels and thermocouples) to recharge the

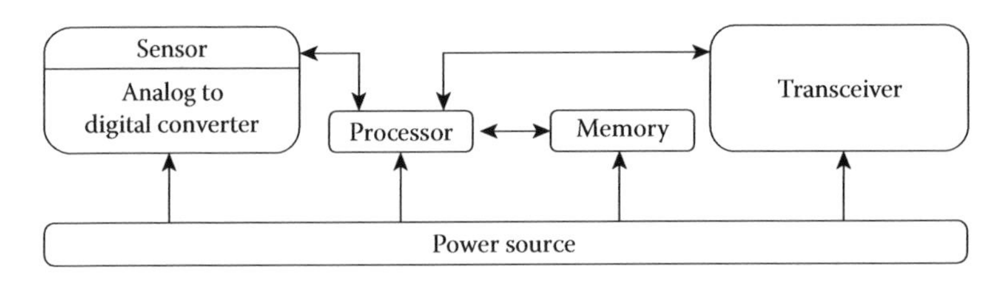

Figure 6.3 The structure of a wireless multimedia sensor network node. (From Sohraby, K.: *Wireless sensor networks: technology, protocols, and applications*. 2007. Copyright Wiley-VCH Verlag GmbH & Co. KGaA.)

node's power supply. However, the five components discussed are crucial for a node to be considered a WMSN node, whereas the other components are optional [2].

6.1.2 WMSN Protocol Stack

The sensor nodes are often scattered in a sensor field. Each sensor node has the functionality to collect and send data to the sink node and end users. Sensed data are sent to the end user by a multihop architecture through the sink node as shown in Figure 6.4. The sink node may communicate with the end user via Internet or satellite [4].

The protocol stack applied by the sink and the sensor nodes is given in Figure 6.4. The protocol stack of WSNs or WMSNs consists of the application layer, transport layer, network layer, data link layer, and physical layer. A brief account of every layer will be listed in the following sections.

The protocol layers can be viewed as a set of management planes across each layer. Each layer includes power, connection, and task management planes [3].

- The *power management plane* is responsible for managing the power level of a sensor node for processing, sensing, transmission, and reception, which can be applied by employing efficient power management schemes at different protocol layers.

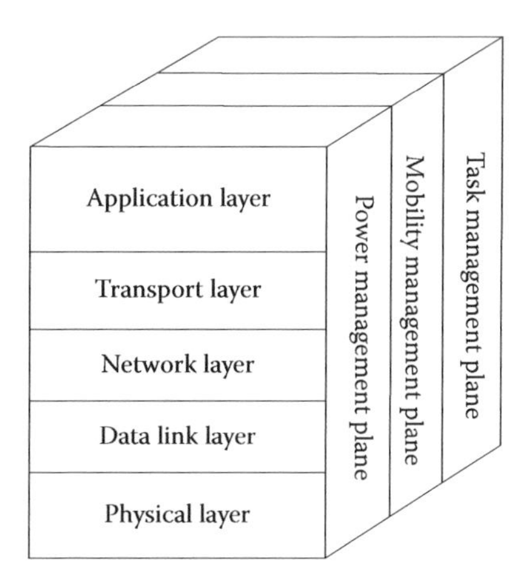

Figure 6.4 The protocol stack for a wireless multimedia sensor network. (From Akyildiz, I.F. et al., *IEEE Communications Magazine*, 40(8), 102–114, © 2002 IEEE.)

- The *connection management plane* handles the configuration and reconfiguration of sensor nodes to establish and maintain the connectivity of a network in the case of node deployment and topology change due to node addition, node failure, or node movement.
- The *task management plane* is in charge of assigning tasks among sensor nodes in a sensing area to improve energy efficiency and extend network lifetime.

6.1.2.1 Application Layer The application layer is a program or a group of programs that is prepared for the end user [2]. There are two types of services: global and local. Global services make direct contributions to the mission objective for a program. For example, a node's decision to track a vehicle would be a global service. Local services are the services that the node provides to itself. For instance, determining the quality of service, performing authentication, and determining communication partners are all examples of local services. Generally, for WMSNs, the application layer determines the system level behavior of a node [2,3].

6.1.2.2 Transport Layer The transport layer is responsible for passing data to the application layer in such a manner as to relieve the application layer of responsibility for cost-effective transmission and connection reliability [2]. The transport layer manages connections, provides congestion and flow control, and maintains connection reliability. Examples of reliable transport protocols used for wireless communications include the User Datagram Protocol (UDP) and Transmission Control Protocol (TCP). It should be noted that in some WSNs a formal transport layer may not exist [2].

6.1.2.3 Network Layer The network layer establishes, maintains, and terminates network connections [2]. The protocol in the network layer is assigned to route, address, and possibly task the node roles if nodes can perform multiple roles, for example, as sensors or sinks. Examples of routing protocols for ad hoc multihop WSNs include the Destination-Sequenced Distance Vector, Ad Hoc On-Demand Distance Vector Routing Protocol, and Temporally Ordered Routing Algorithm [4].

6.1.2.4 Data Link Layer The data link layer is divided into two sublayers: the logical link control (LLC) layer and the media access control (MAC) layer. The LLC layer is assigned to provide a reliable, error-free link between the MAC and network layers. The MAC layer is primarily assigned to minimize collisions in the medium [3].

The MAC layer minimizes packet collisions using techniques such as carrier sensing multiple access and by multiplexing wireless data from different sensor nodes using time division multiple access, frequency division multiple access, or code division multiple access. In addition to minimizing collisions, the MAC layer may also provide some link reliability services, such as cyclical redundancy checking and unique addressing [3].

6.1.2.5 Physical Layer The physical layer is accountable for the transmission and reception of data across wireless and possibly wired channels [1]. For sensor nodes, the physical layer is essentially the Radio Frequency (RF) section of the node. However, a sink node might be connected to more than one medium. For example, the sink node may be connected to the Internet, as well as wirelessly to sensor nodes. In this instance, the sink's physical layer would need to include an RF section as well as a wired modem section.

An effective sensor network is the one that achieves or exceeds the specific performance requirements for its particular application. To achieve specific performance requirements, any proposed scheme design may need to be optimized within and across the node architecture layers.

6.1.3 WMSN Applications

There are several applications for wireless multimedia sensors, in many areas, from military and homeland security, to medical, to increasing efficiency of businesses. The generation of sensors provides tremendous flexibility to applications developers.

Sensors can run while unconnected to the Internet; they can continue to sense, filter, and process local information. Smart sensors can join context to sensed data; for instance, depending on what the item is, sensors can determine if it is broken or spoiled.

There is a range of diverse applications that lend themselves to WMSNs. We will categorize them here into five areas: military, environment, industrial, health, and home and commercial areas [5].

6.1.3.1 Military Applications WMSNs are suitable for use in many military applications, including command and control applications, surveillance and reconnaissance of enemy targets, monitoring for deployment of biological and chemical weapons, and ordinance tracking. The ability of WMSN nodes to be quickly deployed and organize themselves autonomously into functional, fault-tolerant networks makes WMSN technology ideal for these and many other military applications. Some examples of such military applications are listed below [5]:

- Monitoring friendly forces, equipment, and ammunition
- Battlefield surveillance
- Reconnaissance of opposing forces and terrain
- Targeting guidance
- Battle damage assessment
- Nuclear, biological, and chemical attack detection and reconnaissance

6.1.3.2 Environmental Applications Environmental applications for WMSNs include monitoring the movement of wildlife, such as animals and birds, monitoring factors that affect the expansion of crops, exploring inhospitable parts of the earth that remain unexplored, detecting and tracking forest fires, detecting and monitoring the headways of floods, and measuring levels of pollution across distributed locations [2].

6.1.3.3 Industrial Applications In industry, WMSNs can be applied for observing manufacturing processes or the condition of manufacturing equipment; wireless sensors can be attached to production and assembly lines to monitor and control production processes [1].

6.1.3.4 Medical Applications One of the most important applications for WMSNs is in the health-care field. WMSNs provide the potential to offer vulnerable populations a greater level of independence in their lives. Potential applications comprise integrated control of household devices from a single controller, remote patient monitoring, better real-time monitoring of patients in hospitals, and tracking staff and equipment in hospitals [2]. Other examples of WMSN medical applications are as follows:

- Providing interfaces for the disabled
- Integrated patient monitoring
- Administration in hospitals
- Video monitoring of human physiological data
- Tracking and monitoring doctors and patients inside a hospital

6.1.3.5 WMSN Home and Commercial Automation Applications Home and commercial automation technology has been a subject of research for many years. The potential applications for home automation technology are diverse, including multimedia entertainment, automatic houseplant watering, domestic robots, home security, and energy saving [5].

6.1.4 WMSN Research Areas

WMSNs are currently receiving significant attention due to their unlimited potential. Practically, WMSNs encounter numerous problems. The following section will present some areas of research for WMSNs.

6.1.4.1 Real Time in WMSNs WMSNs transact with real-world environments. In many cases, sensed data must be delivered within time constraints so that actions can be taken. There are a few current solutions for meeting real-time requirements in WMSNs [6]. The majority of protocols either ignore real-time processes or simply attempt to process as fast as possible and presume that this speed is enough to meet deadlines.

Researchers have performed several trials in this field to suggest new ideas to solve this problem, namely Real-Time Communication Architecture for Large-Scale Wireless Sensor Networks (RAP) [7] and a Real-Time Routing Protocol for Sensor Networks (SPEED) [8].

Working with real-time protocols usually requires differentiated services. For example, routing solutions need to uphold different classes of traffic: guarantees for the important traffic and less support for unimportant traffic [6]. It is important to design not only real-time protocols for WMSNs but also analysis techniques associated with these protocols.

6.1.4.2 Power Management in WMSNs Limited processor bandwidth and small memory are two arguable constraints in sensor networks, which will gradually disappear with the development of fabrication techniques by manufacturers [9]. However, energy constraints are unlikely to be solved in the near future, due to slow progress in improving battery capacity. Furthermore, the untended nature of sensor nodes and dangerous sensing environments prohibit battery replacement as a logical solution.

By contrast, the monitoring functions of many sensor network applications demand a longer lifetime; therefore, providing a form of energy-efficient monitoring for a surveillance area is a notable research issue [9].

6.1.4.3 Programming Abstractions in WMSNs A key growth area for WMSNs is promoting the level of abstraction for programmers. Currently, developers deal with too many low-level details regarding sensing and node-to-node communication [10]. As an example, they typically deal with sensing data, fusing data, and moving data. If the level of abstraction were raised to consider aggregate behaviors, application functionality, and direct support for scaling issues, then productivity would increase. There have been a few attempts at this type of research; one of these trials was the environment-based abstraction EnviroTrack [11].

Programming abstractions for WMSNs will likely remain rare. Rather, a number of solutions will emerge, each more suited to a certain domain. Research in this area is important in order to expand the development and deployment of WMSNs by general programmers, not by WMSN engineering specialists [11].

6.1.4.4 Security and Privacy in WMSNs WMSNs are limited in their energy, computation, and communication capabilities. In contrast to traditional networks, sensor nodes may be deployed in easily accessible areas, presenting a risk of vandalism. Sensor networks react closely with their physical environment and with people, arguing additional security problems [12].

Because of the above-mentioned problems, the current security mechanisms are inadequate for WMSNs. The new constraints create new research challenges in the areas of key establishment, secrecy and authentication, robustness, stand against denial-of-service (Dos) attacks, secure routing, and node capture [12].

To establish a secure scheme, security should be integrated into every component, since components designed without security in mind can become a point of attack. Eventually, security is a difficult challenge for any system. The strict resource constraints of WMSNs make any computer security for these systems even more challenging [12].

6.1.5 Scope

The work presented in this chapter can be classified under the umbrella of the following areas:

- Real time
- Power management (energy consumption)
- Security and privacy

6.2 Security in WMSNs

Since computers were invented, there has been a need for a strategy to keep data on these computers secured; scientists termed this strategy computer security. With the advent of distributed systems connected by networks across the Internet, there arose a need to keep data safe between these networks; this was called Internet security. With the appearance of WSNs and WMSNs in the research and practical fields, security in these networks has become a necessity. Security here is concerned with how to protect sensed data against unauthorized access and modification and to ensure the availability of network communication and services despite malignant activities.

6.2.1 Security Properties

To consider any scheme secured, there are general security properties that must be satisfied; these properties include authentication, confidentiality, data integrity, and availability [13]. The following subsections will discuss these categories in more detail.

6.2.1.1 Authentication The authentication process is concerned with verifying the identity of an entity (peer authentication) or verifying the source of a message (data authentication) [13].

- In *peer authentication*, the identity of an entity is confirmed to grant physical access to a building or electronic access to a service [13].
- In *data authentication*, the data received are validated to ensure that the data were sent from the entity claimed. The objective of data authentication is to prevent third parties from inserting data into preexisting communications with authenticated entities [13].

The most common approach to peer authentication is providing username and password combinations, smart cards, or biometric data to authenticate users, before providing them with physical access to a building or services. In data authentication, a secret key is used, only known by the sender and the receiver, to encrypt and decrypt data [13].

6.2.1.2 Confidentiality Confidentiality is concerned with keeping data secret from third parties not authorized to view the contents of communications. A system that implements confidentiality must protect data from direct and indirect interpretation by unauthorized entities [13].

Direct interpretation involves unauthorized entities intercepting and viewing data. Indirect interpretation involves unauthorized entities viewing communication traffic patterns and deriving the contents of the communications, through traffic analysis [13].

The primary approach for providing confidentiality is encryption. There are two categories of encryption used to provide confidentiality: symmetric key cryptography [14] and public key cryptography [15].

6.2.1.3 Integrity Data integrity is concerned with preventing the unauthorized modification of data, without detection by a legitimate user.

There are three established approaches for verifying that a message has not been tampered with, namely, message encryption, MACs [16], and hash functions [17].

6.2.1.4 Availability The availability of an item is its state of readiness to execute a required function at a given instant of time or at any instant of time within a given time interval, assuming that the exterior resources, if required, are provided [13].

In the security domain, service availability research focuses on the protection of service-providing systems from attackers that attempt to overwhelm the resources of these systems in an attempt to either permanently remove the service from availability or to sufficiently degrade a service to intermittently remove it from availability. Threats to the availability of services are collectively termed denial-of-service attacks [18].

6.2.2 Security Constraints for WMSNs

Section 6.2.1 discussed general security properties. As already stated, for any system to be considered secure, it must give consideration to the properties of security (i.e., authentication, confidentiality, data integrity, and availability). However, due to the extremely resource-constrained nature of sensor nodes, their security properties may be slightly different compared to those of conventional networks. Revisiting the security properties from a WMSN perspective yields the following [19]:

- *Authentication*: As a WMSN communicates, critical data participates in a number of decision-making procedures. The sink node needs to ensure that the data used in any decision-making process are produced from the correct source node. Similarly, authentication is substantial during the exchange of controlled information in the network.
- *Integrity*: During its journey across the network, data can be changed by adversaries. Data loss or deterioration can also occur without the presence of a malicious node, due to the severe communication environment. Data integrity assures that information is not changed during conveyance, either due to malicious intent or by coincidence.

- *Confidentiality*: Applications such as monitoring of information, industrial secrets, and key distribution depend on confidentiality. The standard approach for keeping confidentiality is through the use of encryption techniques.
- *Availability*: Sensor nodes may use up their battery power due to extreme computation or communication and become unavailable. Security requirements not only influence the operation of the network but are also highly important in maintaining the availability of the network.

Because of the resource-constrained nature of WMSNs, other properties and limitations should be taken into consideration while designing security scheme such as the following [19]:

- *Freshness*: Even if confidentiality and data integrity are indisputable, we also need to ensure the freshness of each message. Freshness of sensed data confirms that the data are recent, and it ensures that no old messages have been replayed. To ensure that no old messages are replayed, a time stamp can be added to the packet.
- *Scalability*: Scalability is required in most WMSN applications, as the number of sensor nodes to be deployed is in the order of hundreds, thousands, or more. The protocols must be sufficiently scalable to reply and operate with such a large number of sensor nodes.
- *Unattended operation*: In many applications, there is no human interference after sensor nodes are deployed. Thus, the nodes are responsible for reconfiguration in case of any modifications. This constrains any security scheme to be as energy sufficient as possible.
- *Untethered*: The sensor nodes are not attached to any energy source. They have only a limited source of energy, which should be optimally used for data processing and communication. To attain optimal energy use, communication must be reduced as much as possible.
- *Limited memory and storage space*: A sensor is a tiny device with only a small amount of memory and storage space for the code. To build an effective security scheme, it is necessary to limit the code size of the security algorithm.

6.2.3 *Attacks on WMSNs*

Security in any network system does not involve only one or two layers, but rather needs to be viewed across all layers as a whole. The security issues for a conventional network differ greatly from the security issues in WMSNs because of the extremely limited resources available in sensor nodes. Attacks in a WMSN can be categorized into two types, passive and active attacks [15].

- In a passive attack, the attacker can get information from the WMSN by eavesdropping on wireless communications and trying to steal confidential data.
- In an active attack, the attacker can get information from the WMSN by spoofing or altering packets to breach authenticity of communication or by injecting false data to impasse the network.

The prime objective of this research is to stand against eavesdropping, stealing confidential data, and injecting false packets into the network; for example, in terms of military applications, an attacker may cause the following:

- Jamming and/or confusing the network protocols by injecting false packets, which leads to DoS in the whole WMSN network (i.e., violating WMSN availability and integrity)
- Eavesdropping on classified information (i.e., violating WMSN confidentiality)
- Supplying misleading information, for example, enemy movements in the east, when in fact they are in the west (in technical terms, violating WMSN authenticity)

Additionally, in health applications, attacks can lead to fatal errors in the health-care system by providing incorrect physiological measurements to the nurse or doctor on behalf of a patient; an offender may cause a potentially fatal diagnosis and treatment to be performed on the patient (i.e., a violation of WMSN integrity and authenticity).

The following subsections will explain the attacks on the WMSN from the perspective of the WMSN layer stack and how to defend against them. Particular emphasis will be placed on attacks on the upper layers (application and transport layer), which are the main focus of this chapter.

6.2.4 Attacks on WMSN Layers

Attacks can also happen on different layers of the WMSN stack layers. The standard layered architecture of the communication protocol for WSNs was introduced in Section 6.1.2 in detail. Attacks and their security solution approaches in different layers with respect to the WMSN layer stack are summarized [20] in Table 6.1. This table provides a classification of the various security attacks on each layer of a WMSN.

6.2.4.1 Transport Layer Attacks The focus of transport layer attacks is to exploit communication protocols that use connection-oriented communications and maintain connection information. The main transport layer attacks against WMSNs include desynchronization attacks and flooding attacks.

Table 6.1 Wireless Multimedia Sensor Network Layer Attacks and Security Approaches

LAYER	POSSIBLE ATTACKS	SECURITY APPROACHES
Application layer	Path-based denial-of-service attack [21] Node reprogramming attacks [19]	Cryptographic approach Authentication
Transport layer	Desynchronization attack [22] Flooding attack [22]	Authentication Complex puzzles [22]
Network layer	Sybil attack [23] Sinkhole attack [23] Wormhole attack [23] Hello flooding attack [23]	Three-way handshake [24] Authentication Cryptographic approach
Data link layer	Collision attack [12] Interrogation attack [12] Denial-of-sleep attacks [25]	Spread spectrum techniques [26] Error correcting codes [27] Rate control mechanisms [27]
Physical layer	Node-tampering attack [19] Jamming and interception attack [19]	Spread spectrum techniques MAC layer admission [28] Tamper proofing (camouflaging nodes) [19] Directional antenna for access restriction [29]

- *Desynchronization attack*: An attacker objects active communications and modifies or fakes the parameters of captured messages, such as control flags and sequence numbers. The modified or faked messages are sent back into an active communication stream between two nodes [22]. Consequently, when modified or faked messages arrive at their respective destinations, they are rejected as out of sequence or as corrupted, leading the sender to resend messages and wasting energy and network bandwidth. The encryption of message headers or the whole message with authentication can ban attackers from modifying existing messages and creating fake packets. Moreover, anti-replay mechanisms [22] can prevent false messages from being inserted into false communication streams undetected.
- *Flooding attack*: The target of the attacker in a flooding attack is a network employing connection-oriented communication protocols. The attacker requests a connection from a node in the WSN; the node detains space in its open connection buffer and sends a synchronization acknowledgment. After a period of time has elapsed, a time-out counter expires, causing the victim to clear its open connection buffer. However, an attacker may repeatedly order a number of connections and leave them half open, thereby exhausting the victim's connection buffer and preventing false connection requests for the duration of the attack [22].

One approach for protecting against flooding attacks in WMSNs demands connecting nodes to complete a complex puzzle [22] before a node reserves connection space. Nevertheless, the use of complex puzzles requires all nodes in the WMSN to have additional hardware to solve complex puzzles. Moreover, each connection attempt will incur false hosts, additional energy, and processing costs.

There are two major protocols in the transport layer, TCP and UDP [3]. In TCP, the attacker creates a large number of half-opened TCP connections with a receiver node, but never finishes the handshake to fully open up the connection, leading to a drain in energy.

Another type of attack in TCP is session hijack. Here, the attacker mimics the victim's Internet Protocol (IP) address, determines the correct sequence number that is expected by the target, and then executes a DoS attack on the victim node [14]. To hijack a session over UDP, the attacker takes the same action as in TCP, except that UDP attackers do not need to concern themselves with the overhead of treating sequence numbers and other TCP control fields. Because UDP is connectionless, phasing into a session without being detected is much easier than in TCP [3].

In this chapter, we concentrate on UDP and how to secure it using authentication mechanisms and encryption techniques, which both stand against the above-mentioned attacks; however, flood attacks will not be included as they require special hardware.

6.2.4.2 Application Layer Attacks Attacks targeting the application layer of WSNs focus either on a weakness in application software specific to a particular WSN or on more general inherent weaknesses in the application layer of WSNs. The most popular forms of application layer attacks include path-based DoS attacks and node reprogramming attacks.

- A *path-based DoS attack* is an attack on the reliability of the WMSN network. An attacker floods counterfeit or replayed packets along a multihop, end-to-end routing path. There are a number of methods proposed for defending against path-based DoS attacks; the primary response of most defense approaches is to detect and remove spurious packets along a communication path. There are three generic defense approaches against such attacks [21]:
    - Each node along a communication path shares a secret key with the sender. The sender generates authentication and integrity material for each key/node and appends it to each packet.
    - A modified approach to that discussed in (1) involves a node storing a path key for every potential path in a WMSN; if any node in the network is subverted, an attacker can flood a whole communication path.
    - Rate control mechanisms can also be applied to each node, limiting the amount of replayed packets accepted from any one node. However, due to the nature of WMSNs, certain nodes—such as nodes directly around a coordinator or router nodes—have different packet rates.

- *Node reprogramming attack*: Due to the nature of many WMSN applications, where nodes are located in inaccessible and remote locations, it is desirable to remotely and wirelessly update node software. The process of updating node software is referred to as code dissemination. There are numerous approaches and protocols for disseminating software, such as the approach adopted by TinyOS, which is called Deluge [30]. In the Deluge approach, nodes periodically send advertisements containing their software version. Secure methods for reprogramming nodes have emerged; one such scheme is called Seluge [31]—a secure extension of the Deluge approach.

From the protocol perspective, the application layer contained user data, and it supported many protocols such as HTTP, SMTP, RTP, and FTP [32], which have vulnerabilities and provide access points for attackers. As mentioned earlier, preprogramming node attacks (malicious code attacks), such as viruses, worms, spyware, and Trojan horses [32], can attack both operating systems and user applications. These malicious programs can usually spread themselves through the network and cause the WSN node and networks to slow down or even become damaged.

In the current research, the focus is on the first solution presented in path-based DoS attack, thereby using the cryptographic approach and adding the MAC authentication mechanism. This research will further focus on Real-Time Transfer Protocol (RTP) as it is commonly used in data streaming.

6.2.5 *Attacks and Scope of Work for the Proposed Security Scheme*

The preceding sections have illustrated attacks that might occur in every layer in the WMSN layer stack. The focus of this study is to defend against attacks in the application and transport layers, as long as the attackers ultimately seek the information that exists within these layers. However, security in any network system does not involve only one or two layers, but rather needs to be viewed across all layers as a whole.

The scope of the work has widened to improve the performance of the proposed security scheme in terms of energy consumption and other metrics. Section 6.3 will describe the structure of the proposed security scheme in detail. We kept providing security features in the application and transport layers to stand against attacks within these layers, but some attacks were hard to defend against as they required hardware or a sensor operating system that was beyond the research objective.

The research objective is to propose a security scheme for multimedia streaming using RTP and UDP within the application and transport layers of the WSN protocol stack. The scope of the proposed security scheme and the attacks that the scheme will stand against are set forth in Table 6.2.

6.2.6 Existing Security Frameworks

Few security solutions exist for WSNs. Most current security solutions have been developed to be quite general to fit various platforms and scenarios. The existing security solutions deliver security features

Table 6.2 Scope of Protection Offered by the Proposed Security Scheme against Wireless Multimedia Sensor Network Attacks

	ATTACKS THAT THE PROPOSED SECURITY SCHEME WILL STAND AGAINST		ATTACKS THAT THE PROPOSED SECURITY SCHEME WILL NOT STAND AGAINST	
LAYER	ATTACK NAME	DEFENSIVE MECHANISM	ATTACK NAME	REASON
Application layer	Path-based DoS	Authentication mechanism		
	Node reprogramming attack (viruses, worms, spywares, and Trojan horses)	Authentication mechanism and encryption technique	Node reprogramming attack (node operating system software update)	Requires nodes with running operating system (TinyOS)
Transport layer	Desynchronization attack	Authentication mechanism	Flooding attack	Require special hardware
Network layer	Hello flooding attack	Authentication mechanism and encryption technique	Sybil attack Sinkhole attack Wormhole attack	Beyond research scope
Data link layer			Collision, interrogation Packet replay Denial-of-sleep attacks	Beyond research scope
Physical layer	Node jamming	DSDV routing	Node tampering	Beyond research scope

at the link and network layers using the TinyOS or Contiki (http://www.contiki-os.org/) operating systems. They did not cover security in the application and transport layers together.

A common problem with all existing security solutions is that they failed to reduce energy consumption and to fit real-time multimedia streaming.

6.2.6.1 TinySec TinySec (http://tinyos.stanford.edu/tinyos-wiki/index.php/TinySec) is a security architecture currently integrated into TinyOS. TinySec was introduced in 2004 and exists in the data link layer. It provides two different modes of operation [33], authenticated encryption or authentication only.

TinySec uses the SkipJack block cipher [33] with the cipher block chaining (CBC) mode of operation. To assure that the ciphertext length is the same as the plaintext, it uses CBC in conjunction with ciphertext stealing (CS). This is usually written as CBC-CS.

6.2.6.2 MiniSec MiniSec (https://sparrow.ece.cmu.edu/group/minisec.html), released in 2007, claims to address and resolve several obstacles with other security architectures proposed for TinyOS. It has two modes of operation, one for unicast and the other for multicast [34]. Both modes suggest using replay protection with synchronized counters or a bloom filter [34], respectively. Like TinySec, MiniSec uses the Skipjack cipher but with offset code book (OCB) as the mode of operation. OCB provides both authenticity and data secrecy while avoiding ciphertext expansion.

6.2.6.3 TinyECC TinyECC (http://discovery.csc.ncsu.edu/software/TinyECC/) was one of the first successful trials to provide public key cryptography in WSNs [35]. It utilizes elliptic curve cryptography, which has a shorter key length than traditional public key cryptology schemes. The shorter key length results in faster computations and less energy consumption and saves on both memory and bandwidth.

6.2.6.4 ContikiSec The first, and currently only, security solution aimed at the Contiki operating system, ContikiSec (http://www.cse.chalmers.se/research/group/dcs/masters/contikisec/) was implemented in 2009 and provides two modes of operation, like TinySec [36]. The first mode provides authenticity and data secrecy, whereas the second provides authenticity only. ContikiSec employs a 128-bit AES as cipher, with OCB for the first mode and CBC-MAC for the second [37].

6.3 Proposed Security Scheme Design

The proposed security scheme in this research built two levels of security. The first level was the encryption of payload data using AES encryption technique [38], while the second level was the generation of the MAC authentication code by using CMAC authentication mechanism [39]. Both levels achieved the security principles (authentication, confidentiality, data integrity, and availability), which were previously introduced in Section 6.2.1.

This section will introduce the design of the proposed security scheme for multimedia streaming in more detail. It will be divided into three sections. The first section will explain the proposed secured packet structure. A diagram of the sending and receiving process for will be outlined in the next section. A proposed scheduling algorithm will then be presented.

The proposed scheduling algorithm led to a significant decrease in the energy consumed by the network for multimedia streaming. At the end of this section, a brief explanation will be given to justify the selection of NS2 for the development of the proposed security scheme.

The significant contribution of this research is the proposed security scheme for multimedia streaming; in addition, an attempt has been made to implement security features with NS2, since NS2 did not support any security features before. The proposed implementation will help other researchers to begin to use NS2 as a security simulator.

6.3.1 Security Scheme Design

The first element in the proposed security scheme design is the packet structure. The packet structure design should take into consideration the two levels of security (in the application and transport layer). It should also maintain the normal functions of the usual structure of the WMSN packet format, presented in Figure 6.5.

The research concentrated on RTP [40] in the application layer with UDP in the transport layer, due to the nature of the applications,

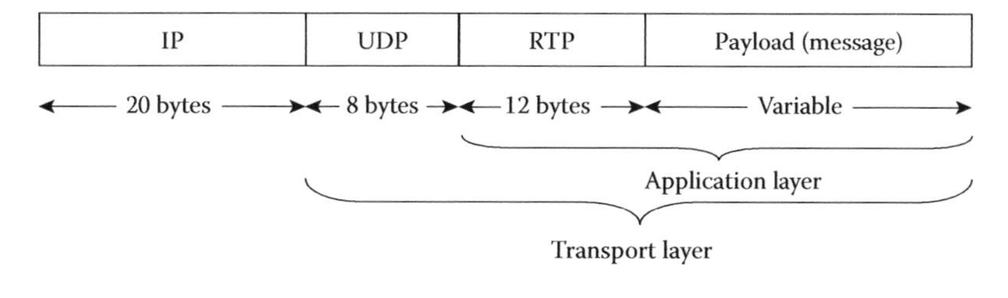

Figure 6.5 Packet format for wireless multimedia sensor networks using RTP and UDP.

such as military and monitoring applications. For example, in military applications, there are two kinds of data transmission. The first type is message transmission, for example, "tank moved in location *X*, *Y*," and the second type is streamed video from sensor cameras, such as video frames.

RTP packets could not be transferred as it was over the WMSN. RTP used UDP to be transferred across the WMSN. To transfer a UDP packet over the WMSN, UDP packets were encapsulated with IP packets [40].

- An IP header contained the information required to route data on the WMSN; it contained the address, the source and destination, and the UDP header.
- The UDP header consisted of four fields, each of which was 2 bytes. The use of the fields "Checksum" and "Source port" was optional. UDP did not guarantee the delivery of payload, which fits the nature of RTP.
- The RTP header had a size of 12 bytes. The RTP was followed by the RTP payload.
- Timestamp and sequence number are the most important fields in the proposed security scheme because they assure the data freshness of the WMSN.

6.3.1.1 Secured Packet Structure Figure 6.6 illustrates the proposed secured packet format for the WMSN. The generated CMAC code depends on UDP, RTP, and the encrypted payload message. The IP header was not included in CMAC calculation, as the research focus was on providing security for the application and transport layers. The generated code would be stored in the MAC header field (4 bytes).

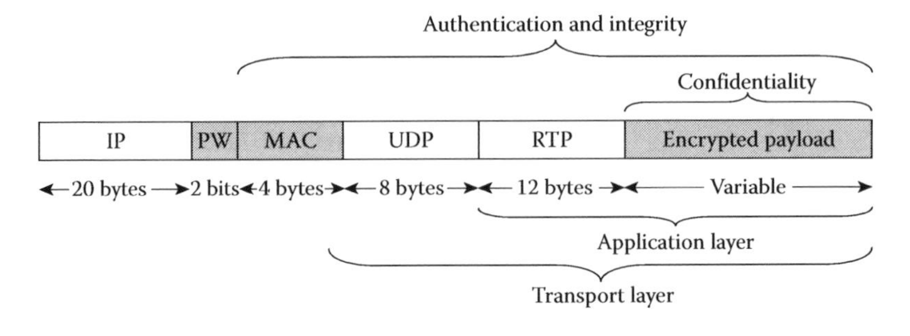

Figure 6.6 Proposed secured packet format for wireless multimedia sensor networks.

The research decision was to select a CMAC authentication mechanism as long as it was developed to fit variable payload data with different sizes. The payload of RTP would carry video frames.

Data freshness was achieved by applying the sequence number and timestamp in the RTP header. A new password (PW) field was added to the proposed secured scheme. The PW field holds 2 bits, and it would contain the values {00, 01, 10, and 11}. The PW values would notify the secured scheme which key password would be used to encrypt and decrypt the payload.

The PW field allowed the proposed security scheme to be more computationally secured against known attacks than any other security solution. The generated MAC field and PW field data in the proposed security scheme were counted as an overhead and will be studied in Section 6.4.9 with the other WMSN security properties.

6.3.2 Proposed Idea for Multimedia Streaming

Multimedia streaming is quite different from message transmission across WMSNs. The difference can be itemized as follows:

- Large data size (frame size varying from 1000 to 3000 bytes).
- Real-time characteristics (frames should be sent and received in a bounded delay).
- Video should be sent at a certain frame rate (standard 24 frames per second).
- The energy consumed by sensors should be taken into consideration while sending and receiving frames.

The above issues made any proposed security mechanism very hard to implement. It encountered some problems in multimedia streaming, because the time between each frame was very small (between 35 and 45 ms), and frame size was more than 1000 bytes to be encrypted, decrypted, and authenticated. The problems can be summarized in three points:

1. *Large frame size*: The size of frames varied from 1000 to 3000 bytes, which would affect encryption, decryption time, and energy. This problem was solved by using a video encoder, which fits the video in the standard frame size for RTP multimedia streaming. The standard frame size will be 1536 bytes, and frame height and width will be 352 × 288 pixels.

   The standard frame size and dimension [41] values were used to calculate the needed time and energy to encrypt, decrypt, and authenticate video frames. This calculation led to a significant savings in the energy consumed by the network for multimedia streaming, while applying the proposed security scheme.
2. *Real-time property* means that the video frames should be sent and received in a bounded delay. Adding security features to real-time video was challenging. We investigated the time needed to encrypt, decrypt, and authenticate a frame size of 1536 bytes in milliseconds. We found that the proposed scheme can be used in real time.
3. *High energy consumption* refers to the energy consumed to send and receive packets, plus the packet processing function. The proposed security features were studied in terms of the amount of consumed energy. Moreover, the research conducted the process of encryption, decryption, and authentication, but it did not consume a lot of energy compared to the sending and receiving function in WMSNs.

Figure 6.7 clarifies the sent function of the proposed security scheme for multimedia streaming. The scenario is as follows:

1. A frame (F) is going to be transmitted by a video camera sensor node.
2. Before sending the frame, a decision must be made by the proposed scheduler whether to send the frame or not.

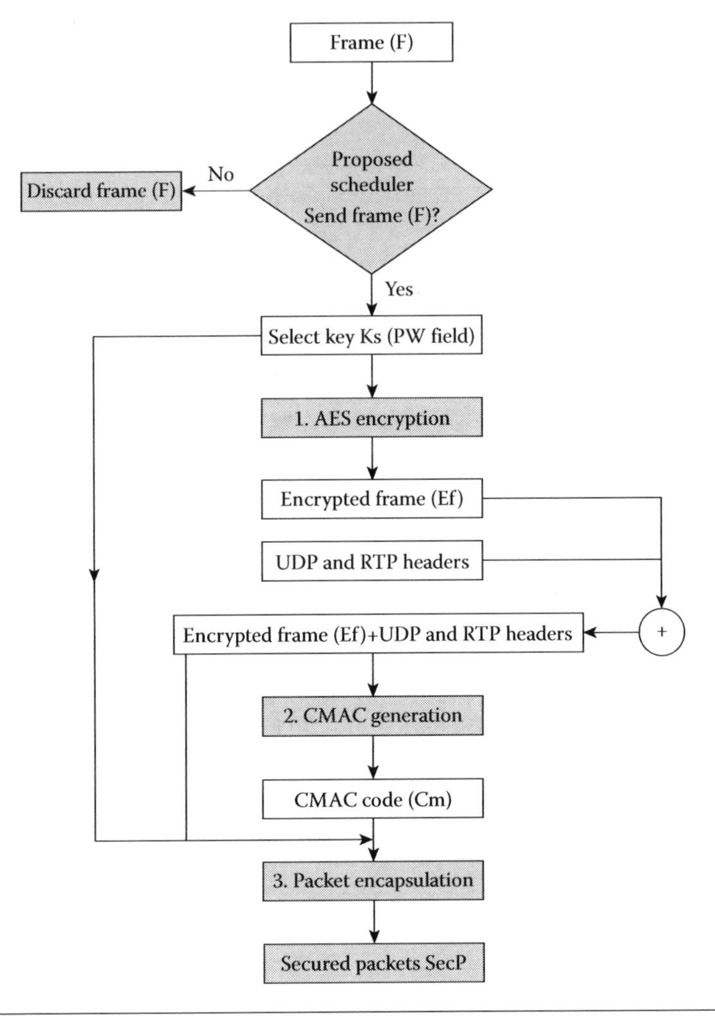

Figure 6.7 Flow chart of the sent function to secure a video frame in wireless multimedia sensor networks.

3. If the scheduler decides to send the frame, then a key will be selected from four different keys that exist in the sent function. If the scheduler decides not to send the frame, no action is taken.
4. The frame will be encrypted with the selected key (Ks), producing an encrypted frame (Ef).
5. The encrypted frame with UDP and RTP headers will be used to generate a CMAC code (Cm).

6. The selected key value, encrypted frame, UDP and RTP headers, and generated CMAC code will be encapsulated together to produce a secured packets (SecP).
7. The secured packet will be sent through the wireless transmission medium.

A frame could be lost for three reasons in the proposed security scheme for multimedia transmission:

1. Due to the encryption process by the sender node, a new frame needs to be sent while the current frame is being encrypted
2. Due to the decryption process in the receiver node, a new frame needs to be displayed while the current frame is being decrypted
3. Due to network traffic congestion

The scheduler proposed by the author led to a significant savings in the network energy consumption for multimedia streaming. The process is introduced below (see also Figure 6.8):

1. Initialize the time of encryption, decryption, and authentication as T_{ENC}, T_{DEC}, and T_{AUTH}.
2. Calculate the time needed to secure a frame $\{T_{SEC} = T_{ENC} + T_{DEC} + T_{AUTH}\}$.
3. Buffer two frames, frame A and frame B, and calculate the time between them as T_{AB}.
4. If $T_{AB} > T_{SEC}$, then pass frame A to the proposed security scheme, or else discard frame A and continue until the end of all video frames.

For instance, if T_{SEC} is equal to 40 ms, and the time between frames A and B is 42 ms, then T_{AB} is greater than T_{SEC}, so frame A will be secured and sent across the network to the receiver node. Otherwise, the frame will be discarded by the sender (the video camera sensor).

The proposed scheduler would decrease network energy consumption, because some frames were going to be dropped by the receiver node due to the decryption and authentication processes.

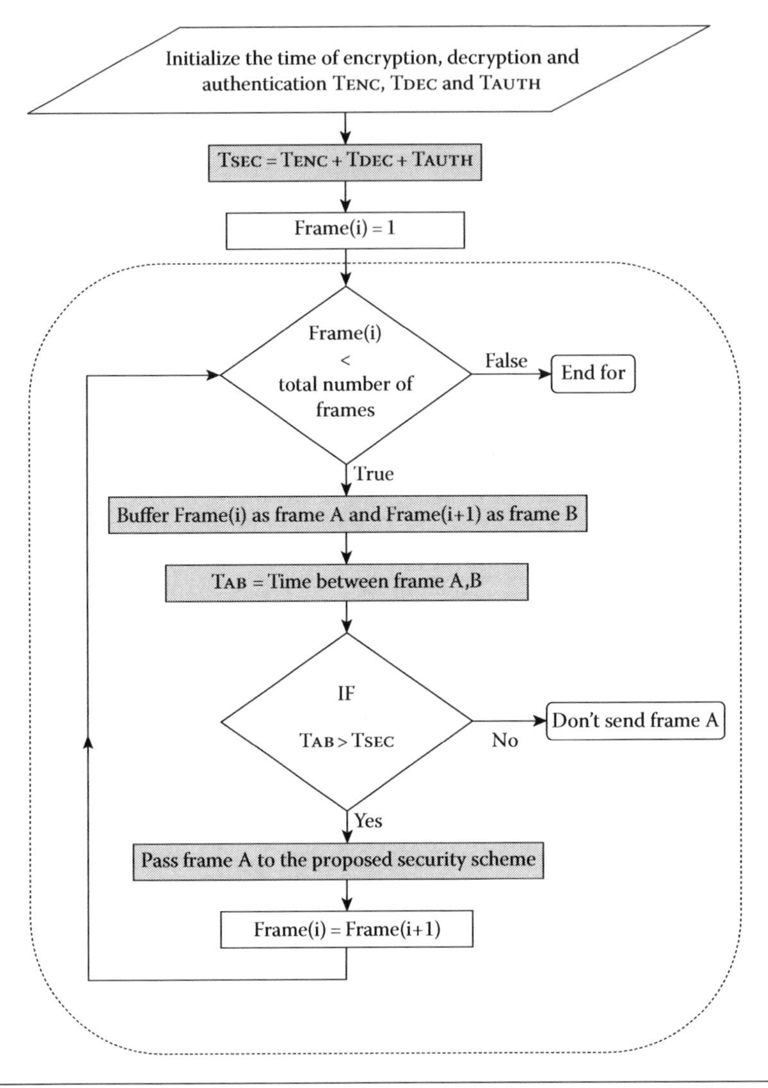

Figure 6.8 Flow chart of the scheduler to secure a video frame in wireless multimedia sensor networks.

Figure 6.9 illustrates the receiving process of the proposed security scheme for multimedia streaming:

1. A secured packets (SecP) is received by the receiver node.
2. The receiver node decapsulates the packet into fields.
3. The receiving process calculates the CMAC code (CCm) by using the encrypted frame (EF) and the UDP and RTP headers.

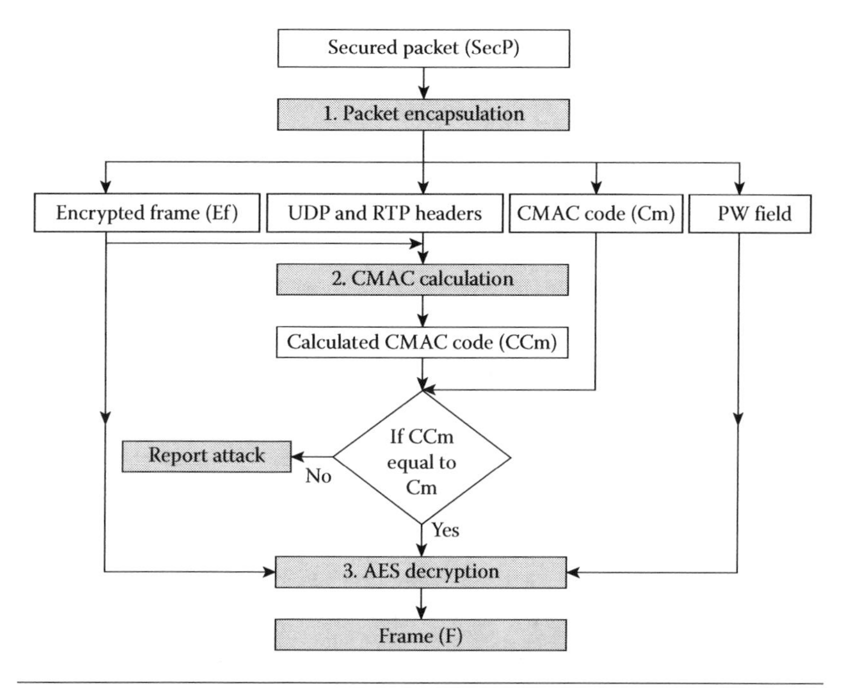

Figure 6.9 Flow chart of the proposed received function to secure a video frame in wireless multimedia sensor networks.

4. If the calculated CMAC code is equal to the sent CMAC code (Cm), then the proposed security scheme sends the encrypted frame to the decryption process; otherwise it reports an attack and shuts down all sensor nodes.
5. The decryption process decrypts the encrypted frame using the AES decryption process and produces frame (F) if the calculated CMAC is equal to the sent CMAC code.

6.3.3 Security Scheme Implementation

There were two possible methods of evaluating the proposed security scheme. The first possibility was to implement the scheme on real sensor nodes in the TinyOS [42] operating system. It was very difficult to choose this method as it required between 80 and 100 sensor motes, which exceeded the research budget. The second method was to implement the proposed security scheme in WMSN simulators; this solution was satisfactory, since it met the research budget and schedule.

A simulator is software that imitates selected parts of the real world and is normally used as a tool for research and development [43]; there are more than 60 simulators for WSNs and WMSNs. Table 6.3 outlines some of the most well-known WSN simulators.

The NS2 simulator was selected from the list of WSN and WMSN simulators. It was decided to implement the proposed security scheme within NS2 for the following reasons:

- It is open-source software.
- It has a large number of protocols available.
- It is used by many researchers around the world.
- It has not supported any security features so far.

Although NS2 has a complicated design, the purpose for using it was to make a valuable contribution to the research community. Appendix 6A will deal with the installation of NS2 and will explain the main components of the simulator. In addition, a detailed structure of the proposed security scheme and its implementation in NS2 will be given.

Appendix 6A will provide further details on how to implement the proposed security scheme for multimedia streaming in NS2.

6.4 Evaluation of the Proposed Security Scheme

Section 6.3 introduced the design of the proposed security scheme for multimedia streaming. In this section, the simulation assumptions, parameters, scenarios, metrics, and results of the proposed security scheme will be presented. It is important to note that all the simulation trials are based on NS2.

6.4.1 Simulation Assumptions

There is a set of assumptions that should first be listed before running the simulations, and they are as follows:

- There is no mobility feature on the nodes. Once nodes are deployed, they cannot be moved or replaced.
- Traffic flow follows a pattern determined by the application. Most of the traffic is assumed to be directed from the nodes to the sink node.

Table 6.3 Comparison of Well-Known Wireless Sensor Network Simulators

SIMULATOR	PROGRAMMING LANGUAGE	GUI	OPEN SOURCE	MAIN FEATURES	LIMITATIONS
QualNet [44]	C/C++	Yes	Commercial	Comprehensive set of advanced wireless modules and user-friendly tools	High-cost annual license
OPNET [44]	C/C++	Yes	Commercial	Uses a hierarchical model to define each characteristic of the system Capable of recording a large set of user-defined results	Scalability problems Expensive tool
J-Sim [45]	Java	Yes	Yes	Ability to simulate the use of sensors for phenomena detection Support for using the simulation code for real hardware sensors	Comparatively complicated to use Unnecessary overhead in the intercommunication model
SENS [44]	C++	No	Yes	Provides very basic network and physical layer support Source codes can be compiled for TinyOS	Simulators do not seem to be developed any further
NS2 [46]	C++	No	Yes	A large number of protocols available publicly Ability to support multiple radio interfaces and multiple channels	Complex configuration Does not run real hardware code

Sources: Eriksson, J., Detailed simulation of heterogeneous wireless sensor networks, Uppsala University, Uppsala, Sweden, 2009.
Kellner, K. et al., Simulation environments for wireless sensor networks, Georg-August University, Gottingen, Germany, 2010.

- All the sensors use integrated circuits that are tamper-resistant. Thus, if a node is captured, the attacker is unable to extract data from the sensor, especially the network keys.
- Traffic generation is based on a fired triggered event in the simulation environment.
- All nodes always have a way to the sink node through other nodes or a direct connection to the sink node.
- Multimedia sensors are more powerful than normal sensors in terms of initial energy and memory.
- The standard frame size for multimedia streaming is 1536 bytes, and screen width and height is 352 × 288 pixels.

6.4.2 Simulation Parameters

To simulate the proposed scheme, the following general NS2 parameters should be set up first:

1. The appearance of the network, which indicates the full view of the topology of the sensor network; this includes the position of nodes with (x, y, and z) coordinates.
2. The internal configuration: Because the simulation is conducted on network traffic, it is important to configure the following:
   a. Which nodes will be the sources?
   b. What is the status of the connections?
   c. What kind of secure connection will be used?
3. The configuration of the layered structure of each node in the network, including the following:
   a. The detailed configuration of network components on a sensor node.
   b. Where to produce the simulation results (the trace file).
   c. Organizing the simulation process.

6.4.2.1 Appearance of the Network for Multimedia Streaming In the multimedia streaming scenario, node positions were generated according to uniform distribution. The monitored area was 1000 × 1000 m. There were 25 nodes deployed. The simulation included from one to four sensor cameras with different x and y coordinates.

The location of x and y coordinates of sensor cameras can be away from the sink node by one to four hops. Figure 6.10 shows an example

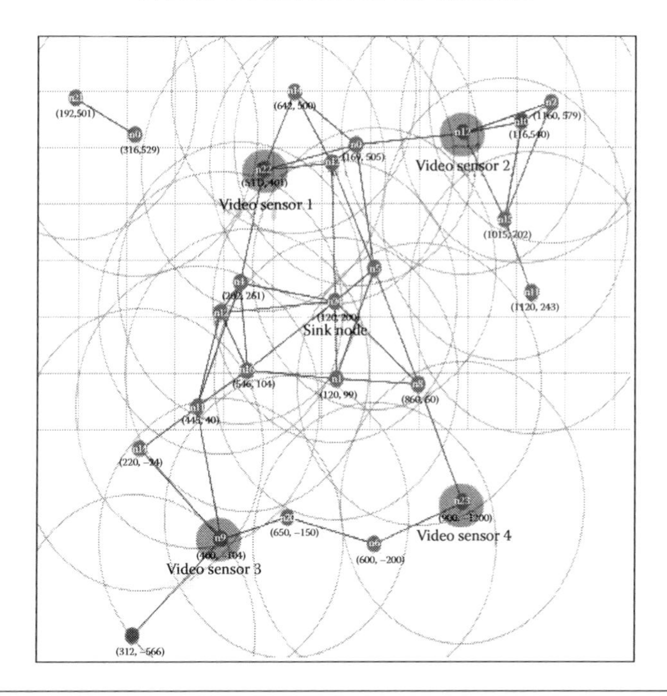

Figure 6.10 Example of one of the network appearance for nodes in 1 km^2 area for multimedia streaming.

of the network appearance of the nodes in a 1 km^2 area with sensor cameras.

6.4.2.2 Internal Configuration Because the simulation is conducted on network traffic, the traffic sources are all deployed nodes except for the sink node. Most of the traffic is assumed to be directed from the nodes to the sink node. The reason for that assumption is that the ultimate goal of the WMSN is to get information from the network. Every node in the simulation scenario is connected to a set of neighboring sensor nodes. There is always a path to the sink node.

6.4.2.3 Layered Structure Configuration of Nodes The layer structure of multimedia streaming is quite different from any other transmission type. Table 6.4 lists the NS2 parameters for a multimedia streaming scheme. WMSNs not only have high traffic flow but also a large frame size for a bounded time. In WMSNs, the unsuccessful reception of large packets is due to the number of hops packets must make to reach the sink; every packet with large multimedia data

Table 6.4 NS2 Simulation Parameters for Multimedia Streaming

SIMULATION PARAMETER	VALUE
MAC layer protocol	IEEE 802.11
Transmission radius	250 M
Data packet size	64 bytes
Data rate	24 frame/s
Traffic type	Multimedia frame
Simulation area	1 km^2
Number of sensor nodes	25
Node initial energy	1000 J
Transmission energy	0.007 J
Reception energy	0.007 J
Simulation times	120 s, 360 s
Routing protocols	AODV

passes from 1 to 20 nodes to reach the sink. The authors suggested using a minimum number of nodes to be deployed in an area 1 km^2. The proposed number of nodes is 25.

The WMSNs had initial energy larger than the WSNs. The initial energy was 1000 J. The times for simulation were 120 and 360 seconds.

The above simulation times were chosen because the simulations had two multimedia streamed videos. The duration of the first video was 20 seconds, and the duration of the second video was 120 seconds. The simulation experiments (Appendix 6A) will give more details about the videos used in such simulations.

6.4.3 Simulation Scenarios

The multimedia streaming scenario was quite different and more complicated than any transmission. Scenarios for multimedia streaming were divided into two scenarios: short multimedia streaming and long multimedia streaming.

The short multimedia streaming duration was approximately 20 seconds. The long multimedia streaming duration was approximately 120 seconds. For both types, simulation experiments should be conducted when there are between one and four videos to be streamed in the network.

The positions of video cameras sensor nodes should be away from the sink node by one hop, two hops, three hops, and four hops;

moreover, some experiments will be executed when video cameras are deployed in random places in the network. The simulation scenarios are quite numerous so as to study the proposed security scheme for multimedia streaming in different conditions and environments.

6.4.4 Simulation Hardware

All simulation scenarios were performed in Windows Server 2003 Enterprise Edition Service Pack 2 (Kansas) using the compiler GCC 2.9.6. All experiments were run on a Dell PowerEdge 2800 tower server (Texas) with an Intel Xeon CPU, 3.40 GHz, and 4 GB of RAM.

The execution of simulation scenarios consumed a large amount of time. The specifications of the machine had no relation to the simulation performance and results. NS2 simulates the specifications of the wireless sensor node hardware. The machine used helped in processing the resulting trace files and videos before and after simulations.

6.4.5 Performance Metrics

In a multimedia streaming scenario, the metrics are the total network energy consumption, network packet delivery ratio, video frames lost due to network congestion or due to the proposed security scheme, and peak signal-to-noise ratio (PSNR).

The main challenges for the proposed security scheme are the energy consumption and the strength of the security to stand against all known attacks without missing the network delivery ratio.

6.4.5.1 Network Delivery Ratio The network delivery ratio is defined as the number of successful received packets during a specific time. Equation 6.1 is used to calculate the network delivery ratio.

$$\text{Network delivery ratio} = \frac{\sum_{i=1}^{n} RP_i}{\text{All Transmitted Packets}} \tag{6.1}$$

where n represents the total number of nodes, RP_i represents the received packets by node i.

6.4.5.2 Average Energy Consumption per Received Packets The average energy consumption per received packet is defined as the total energy consumed by the whole network to deliver control and data packets to the destination during a specific time. Equation 6.2 is employed to calculate that metric.

$$\text{Average energy consumption per received packets} = \frac{\sum_{i=1}^{n} E_i}{\sum_{i=1}^{n} RP_i} \quad (6.2)$$

where n represents the total number of nodes, E_i represents the consumed energy by node i, RP_i represents the received packets by node i.

6.4.5.3 Frame Loss Frame loss is defined as one or more frames of data traveling across the network that fail to reach their destination during a specific time or due to node calculations. Equation 6.3 is drawn upon to calculate the frame loss.

$$\text{Frame Loss} = \text{Dropped frames due to network congestion} + \text{dropped frames due to node calculations} \quad (6.3)$$

6.4.5.4 PSNR PSNR is defined as the normalized average difference between each pixel in the transmitted video and the received video through the network. This is the most commonly used method to measure video quality. Equations 6.4 and 6.5 are used to calculate PSNR.

This measurement has a nonlinear relationship with the subjective video quality. Excellent values for video encoding range from 30 to 50 dB, whereas the acceptable range in wireless transmission settles from 20 to 25 dB.

$$\text{PSNR} = 10 \log_{10} \frac{L^2}{MSE} \quad (6.4)$$

$$MSE = \frac{1}{N * M} \sum_{i=0}^{N-1} \sum_{j=0}^{M-1} \left[X(i,j) - Y(i,j) \right]^2 \quad (6.5)$$

where L represents the maximum value a pixel can take in a video frame, MSE represents the mean square error, X represents the transmitted video frame, Y represents the received video frame, and i,j represents the location of the pixel in a video frame.

6.4.6 Simulation Results

The simulation results of the proposed security scheme for multimedia streaming are included in two sections. The first section will represent the results for short multimedia streaming, while the second one will show the results for the long multimedia streaming. The short and long multimedia streaming have been selected to test the behavior of the proposed security scheme in terms of light and heavy network traffic load.

The metrics to be measured for both sections are the total network energy consumption, network packet delivery ratio, lost video frames due to network congestion or due to the proposed security scheme, and PSNR. The above metrics are commonly used to measure the performance of any proposed work for multimedia streaming [47,48].

6.4.6.1 Simulation Results for Short Multimedia Streaming Streaming short duration video files is very important in many applications such as monitoring and military applications. To classify a movie as short, the standard movie data set classification was used in the proposed work. This classification considers a movie short when its duration is less than 20 seconds and long when its duration is more than 1 minute long, when the wireless transmission medium is used.

All the following simulations result in short multimedia streaming using a movie entitled "Container" with a duration of 20 seconds to be sent through the network (Appendix 6A explains how to prepare a movie for simulation). There were four scenarios in short multimedia streaming with one, two, three, and four video cameras, respectively, placed in the network topology in different positions. The metrics to be measured were the total network energy consumption, network packet delivery ratio, and video frame loss due to network congestion or due to the proposed security scheme.

6.4.6.1.1 The Total Network Energy Consumption The first metric to be measured in the current work was the total network energy consumption. Figures 6.11 through 6.14 represent the energy consumed by the network when there are one, two, three, and four videos streamed through the network, respectively. The *x* axis represents the location of the video cameras, while the *y* axis indicates the network energy consumption in joules.

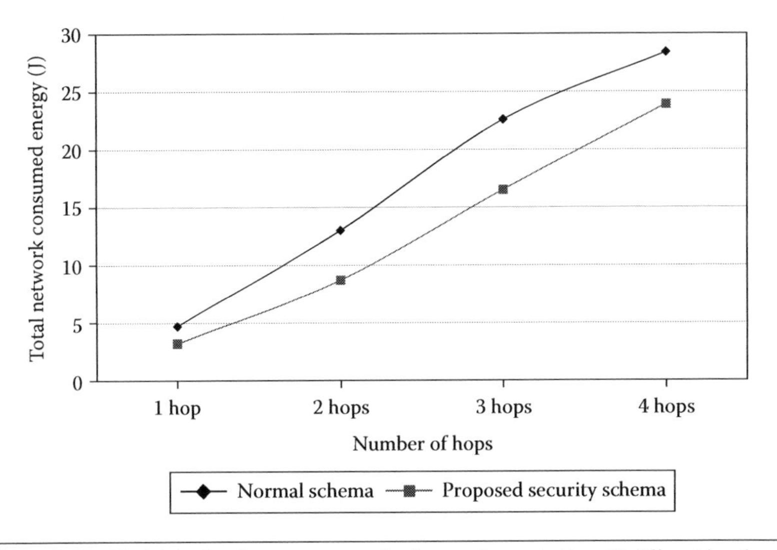

Figure 6.11 The total network energy consumption for one streamed video with different locations.

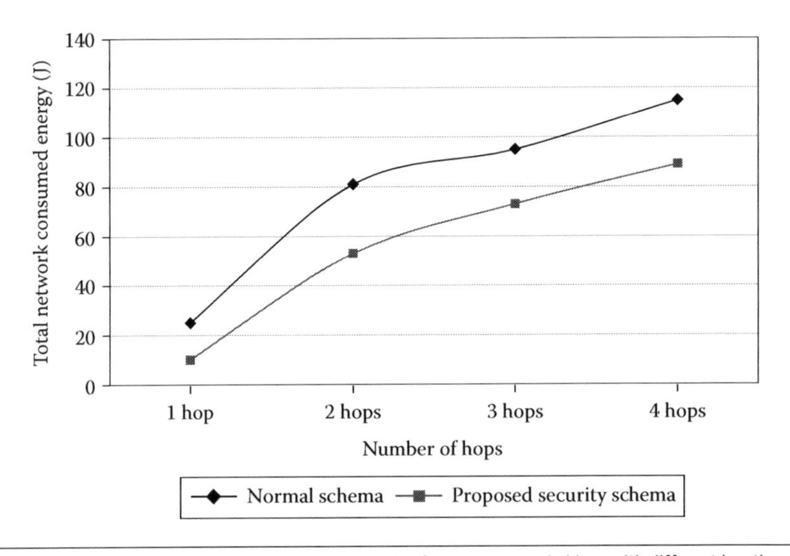

Figure 6.12 The total network energy consumption for two streamed videos with different locations.

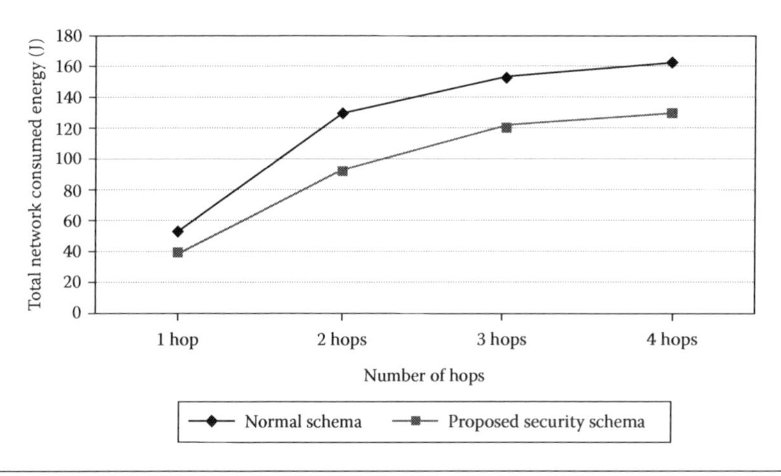

Figure 6.13 The total network energy consumption for three streamed videos with different locations.

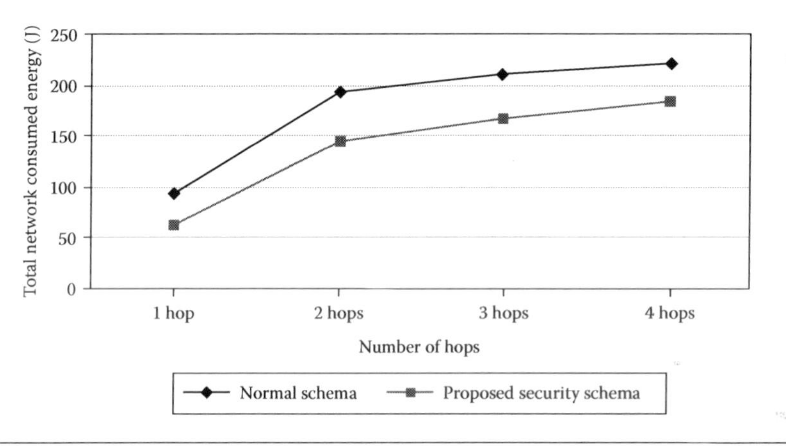

Figure 6.14 The total network energy consumption for four streamed videos with different locations.

Figures 6.11 through 6.14 illustrate that the proposed security scheme achieved less energy consumption than the normal scheme. The proposed scheduler that was presented in Section 6.3.2 achieved this improvement in network energy consumption. The video camera sensor would discard any frame that would not burden the process of encryption, decryption, and authentication before sending it to their destination. This idea conserved a lot of consumed energy, as there were frames that were going to consume a lot of energy during their journey to the sink node. When they reached the sink node, they would be dropped, as it could not burden the process of decryption. The energy consumed during its journey was proportional to the video camera location: the farther away the video camera was located, the more energy the frame consumed to reach the sink node.

The question that might arise here is what the times were when two, three, and four videos were streamed through the network. If the two, three, and four videos were streamed, and there was a pause time between them, then there would not be a problem; the network would handle this situation because there was one video at a time.

Random start times were generated for the videos with a uniform distribution. The uniform distribution would generate two numbers between 0 and 40 for two videos streamed through the network. The value 40 was the result of multiplying the duration of the movie by the number of videos to be streamed. For three videos, the random times to start streaming would be between 0 and 60, and so forth.

The simulations were executed 10 times for every scenario, and the average was calculated from these values to conclude network energy consumption and the other metric.

Note that the behavior of the energy consumption curve for the proposed security scheme was the same in all cases. The proposed security scheme always achieved lower energy consumption than the normal scheme because of the proposed scheduler. The proposed security scheme achieved from 15% to 20% lower energy consumption than the normal scheme.

6.4.6.1.2 The Network Delivery Ratio The second metric to be measured through the proposed work is the total network delivery ratio. Figures 6.15 through 6.18 illustrate the network delivery ratio when there are one, two, three, and four videos streamed through the network, respectively. The x axis indicates the location of the video cameras, whereas the y axis refers to the achieved network delivery ratio as a percentage.

Figure 6.15 shows that the delivery ratio for the proposed security scheme is 81%. The 81% delivery ratio was achieved due to the discarded frames by the proposed scheduler. When the video camera was four hops away from the sink, the delivery ratio of the normal scheme was below 81%, while the proposed secured scheme is

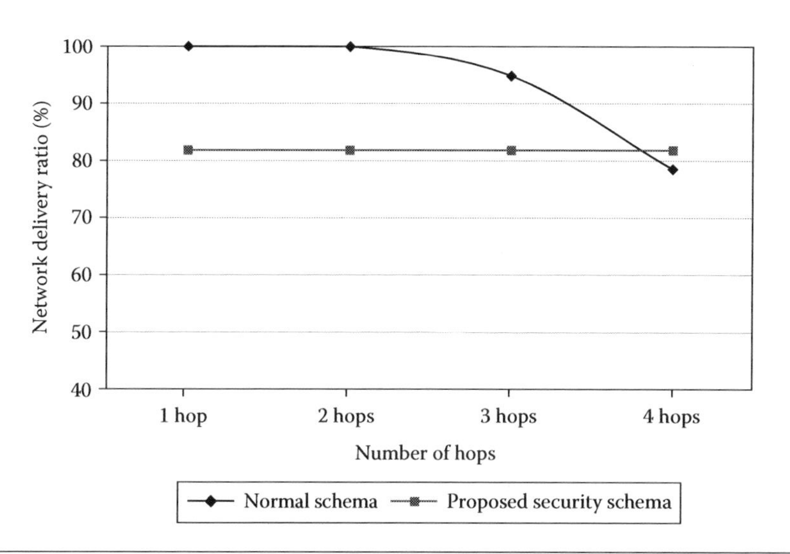

Figure 6.15 The network delivery ratio for one streamed video with different locations.

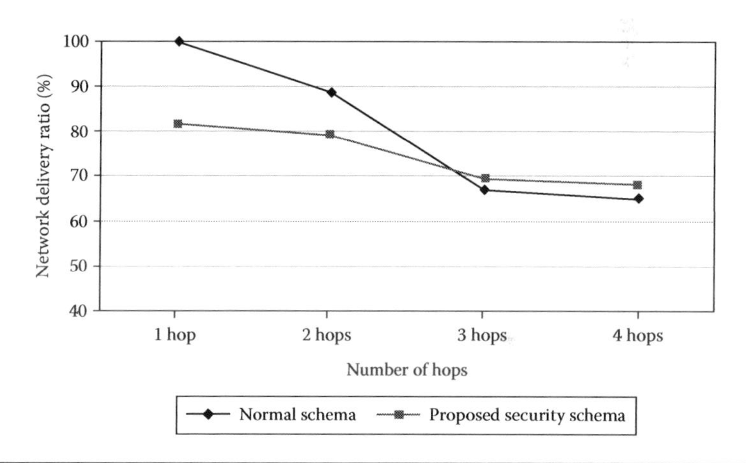

Figure 6.16 The network delivery ratio for two streamed videos with different locations.

constant at 81%. The delivery ratio for the normal scheme was below 81% because of the network traffic congestion.

From Figures 6.16 through 6.18, it is clear that the delivery ratio of the proposed secured scheme was not constant at 81% as when there was one video streamed through the network. The delivery ratio of the proposed security scheme was below 81%. The reason for that phenomenon was due to network traffic congestion as there were two, three, and four video cameras, and their locations were two, three, and four hops away from the sink, respectively.

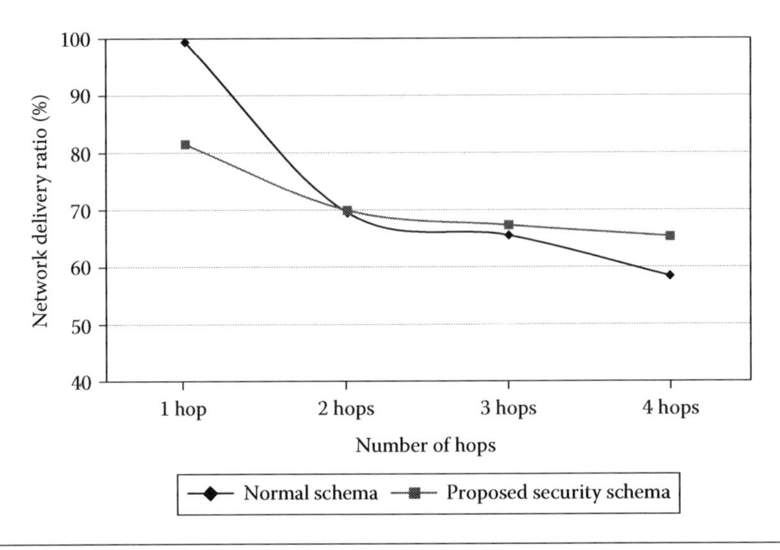

Figure 6.17 The network delivery ratio for three streamed videos with different locations.

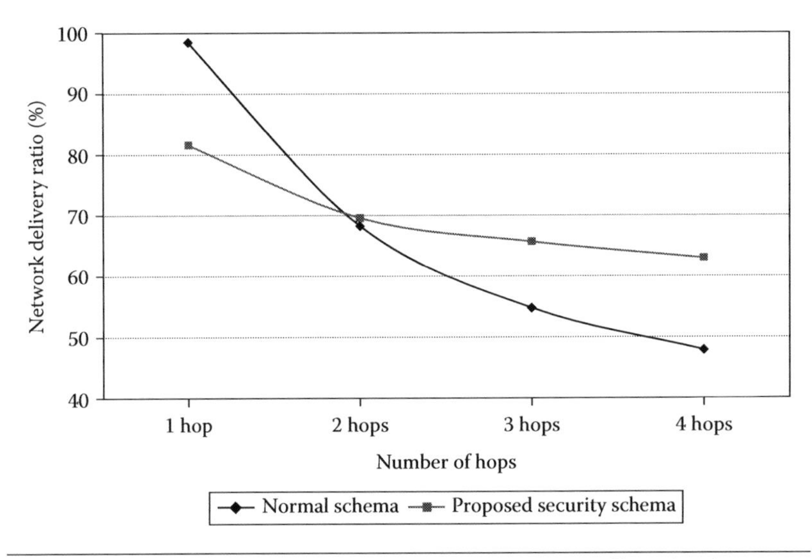

Figure 6.18 The network delivery ratio for four streamed videos with different locations.

For multimedia streaming, the video was considered to be at a good quality for watching when no more 1/3 of frames were being lost. The proposed security scheme and the normal scheme achieved more than 65% delivery ratio as shown in Figures 6.16 and 6.17, which was a good indicator for movie quality. The PSNR will strengthen our indication at the end of the multimedia results section.

Note that, in Figure 6.18 at four hops away from the sink node, the delivery ratio of the normal scheme and the proposed secured scheme was below 65%. This value was not acceptable for video quality. However, this was the worst-case scenario for the whole simulation. Random deployment of video cameras was thus required to draw the final conclusion of the proposed security scheme performance. The PSNR will also have the final decision concerning the quality of the streamed videos.

6.4.6.1.3 Frame Loss The frame loss metric was one of the most important metrics to be measured in the proposed security scheme. This metric presents the reasons for the frames that were being lost. A frame could be lost for two reasons. The first reason for loss was due to network congestion, which occurred with both the proposed security scheme and the normal one, whereas the second reason for loss was the security calculations performed by the proposed security scheme.

Figures 6.19 through 6.22 represent the frame loss when there are one, two, three, and four videos streamed through the network, respectively. The *x* axis indicates the location of the video cameras, whereas the *y* axis represents the number of lost frames.

Figure 6.19 reveals that all the lost frames in the proposed security scheme were due to security process calculations; it did not lose any frames due to network traffic congestion, as it had already dropped

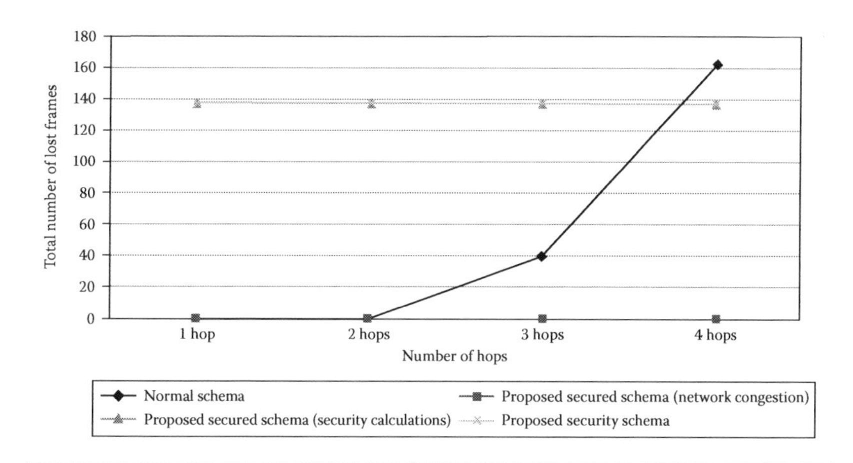

Figure 6.19 The total number of lost frames for one streamed video with different locations.

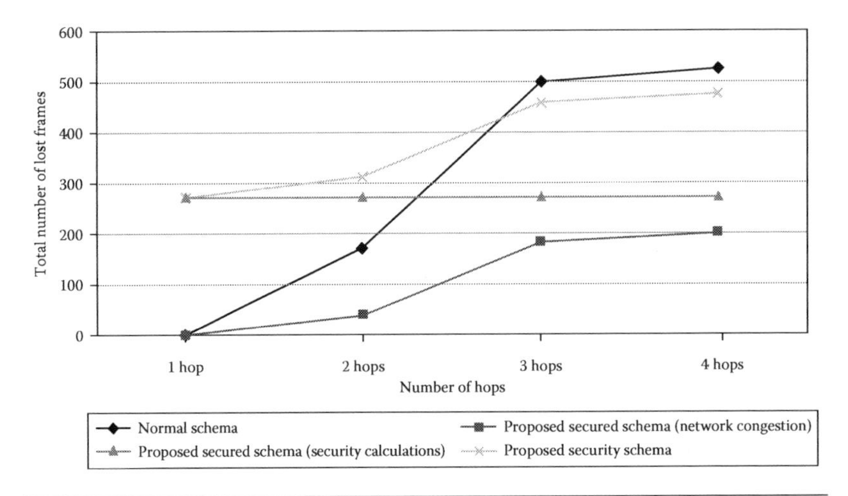

Figure 6.20 The total number of lost frames for two streamed videos with different locations.

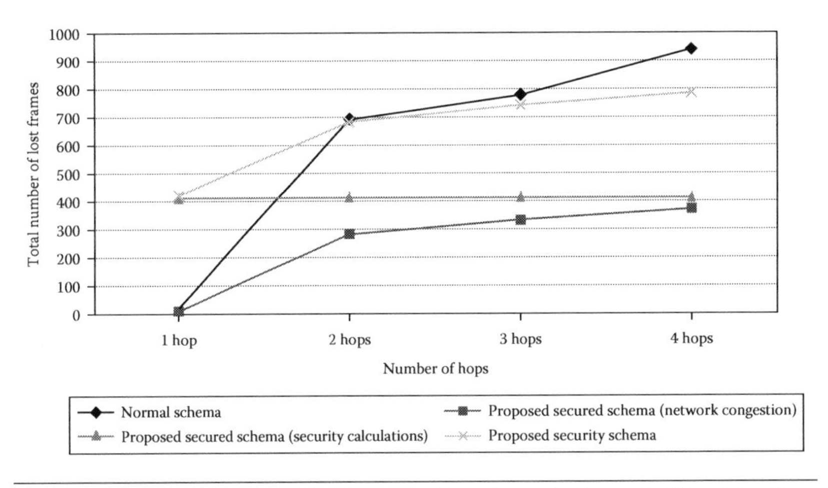

Figure 6.21 The total number of lost frames for three streamed videos with different locations.

the frames that could not burden the security process by the video camera. Hence, it had a lower number of lost frames than the normal scheme; even if the video camera was four hops away from the sink node, all frames were successfully delivered to the sink under the proposed security scheme. In general, the normal scheme had fewer lost frames when the video camera was close to the sink and more lost frames when the video camera was farther away from the sink.

From Figures 6.20 and 6.21, it is apparent that the lost frames of the proposed secured scheme were due to network traffic congestion and the security calculations. Generally, there were two proportional

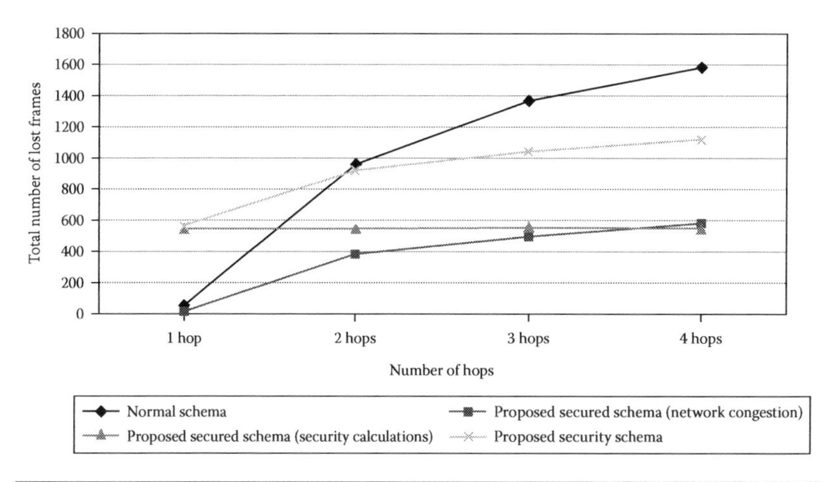

Figure 6.22 The total number of lost frames for four streamed videos with different locations.

relationships; the first one was that the more video cameras existed in the network topology, the more lost frames the network had. The second relationship was that increasing distance from the sink also increased the number of lost frames.

The proposed security scheme was biased to achieve fewer lost frames than the normal scheme when there were more cameras deployed and when they were located farther away from the sink, since the number of sent frames was less than the normal scheme. According to the WMSN application, the video cameras were always deployed far from the sink to monitor the whole region. The choice of deploying video cameras will constitute a design variable controlled by the network engineer in our proposed scheme.

Note that the worst-case scenario happened when four video cameras were deployed away from the sink node by four hops. The network delivery ratio was 62%, and the number of lost frames due to network traffic congestion was equal to the lost frames due to the security calculations. This result poses the question of what the behavior of the proposed security scheme was when the four video cameras were deployed in random locations.

6.4.6.2 Performance Evaluations for Short Multimedia Streaming When the Four Video Cameras Were Deployed in Random Locations Table 6.5 exhibits the random locations for four video cameras when they were deployed in a random fashion. For every combination, the simulations

Table 6.5 Random Locations for Four Video Cameras in Short Multimedia Streaming

	LOCATION OF VIDEO CAMERAS (NUMBER OF HOPS FROM SINK)			
COMBINATION NAME	VIDEO CAMERA 1	VIDEO CAMERA 2	VIDEO CAMERA 3	VIDEO CAMERA 4
Combination 1	1	1	2	2
Combination 2	1	1	3	2
Combination 3	1	1	2	4
Combination 4	1	2	3	4
Combination 5	2	3	3	3
Combination 6	3	3	3	4
Combination 7	3	3	4	4

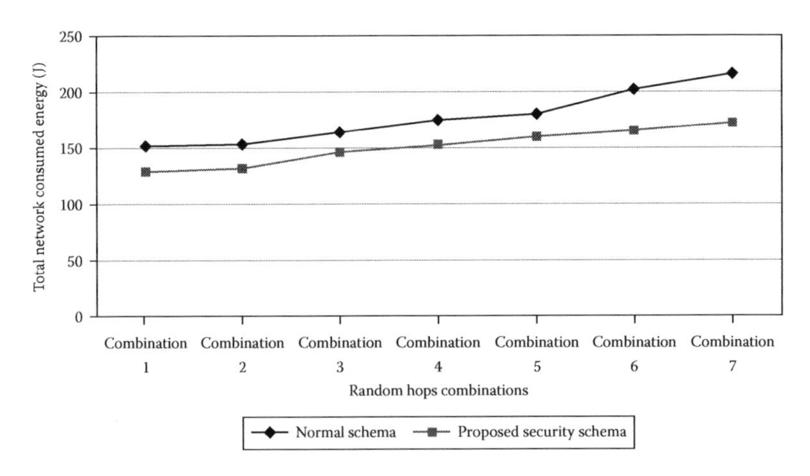

Figure 6.23 The total network energy consumption for four streamed videos with random locations.

were executed 10 times, and the average was calculated from these runs. The metrics to be measured were network energy consumption, network delivery ratio, and total number of lost frames. The random location deployment was a strategy to draw a final conclusion about the performance of the proposed security scheme to figure out its behavior under random deployment.

The first metric to be measured was network energy consumption. Figure 6.23 represents the energy consumed by the network when four videos were streamed from random locations in the network topology. The time to start streaming the four videos was a random time between 0 and 80.

Figure 6.23 illustrates that the proposed security scheme was sufficiently stable with random locations for video cameras according to the energy consumption metric. The proposed security scheme's energy consumption curve always achieved lower energy consumption. There was no doubt that, although the proposed security scheme added security features to the network, the network nonetheless achieved less energy consumption.

It is notable that the proposed security scheme achieved better results when the video cameras were away from the sink by three or four hops. Figure 6.23 illustrates that, in Combinations 6 and 7, the proposed security scheme curve margin was 40 J less than the normal one. It seems that the proposed security scheme had a strong tendency to achieve better results when video cameras were away from sink by more than two hops.

Figure 6.24 demonstrates the network delivery ratio for both the security scheme and the normal scheme when four videos were streamed through the network from random locations. The proposed security scheme achieved a lower delivery ratio when Combinations 1, 2, and 3 were executed, but it achieved a better delivery ratio when Combinations 4, 5, 6, and 7 were executed. The reason for this result was the locations of video cameras; the farther away the video cameras were placed, the better delivery ratio the network had. That is why the number of frames delivered in the proposed scheme was less than the normal one.

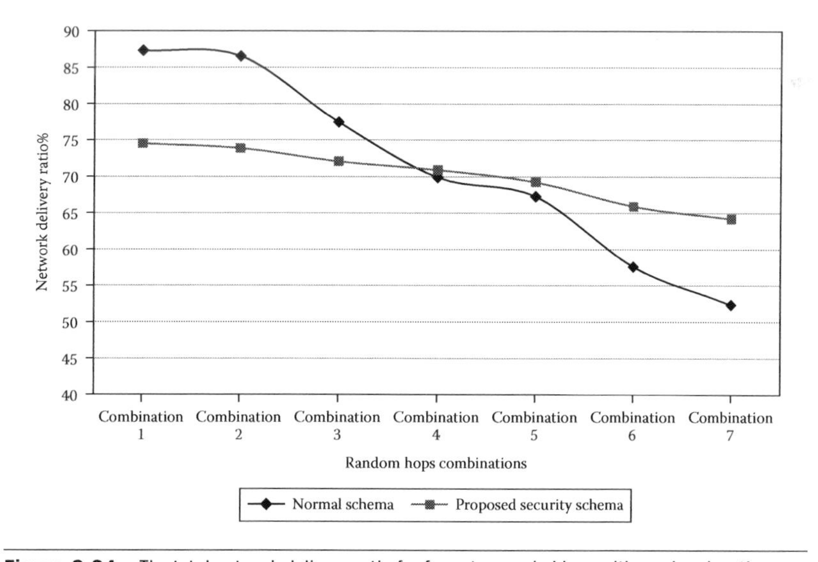

Figure 6.24 The total network delivery ratio for four streamed videos with random locations.

With Combinations 6 and 7, the network delivery ratio of the proposed security scheme was approximately 65%. This ratio was good enough to strengthen our analysis for the case of four video cameras deployed four hops away from the sink; that was the worst-case scenario in the network.

Figure 6.25 shows the network total number of lost frames. It was clear that the proposed security scheme achieved fewer lost frames than the normal one. The frames lost due to the security calculations were always constant, at 18%–20% of the original video. The frames lost due to network congestion in the proposed security scheme were proportional to the locations of video cameras. The farther the video cameras were located from the sink, the more network congestion there would be.

The results of the short multimedia streaming simulation, as presented in Sections 6.4.6.1 and 6.4.6.2, can be summarized as follows:

1. The proposed security scheme achieved lower energy consumption than the normal scheme in all simulation scenarios.
2. The lower energy consumption of the proposed security scheme was achieved because of the proposed scheduler, as already discussed in Section 6.3.2.

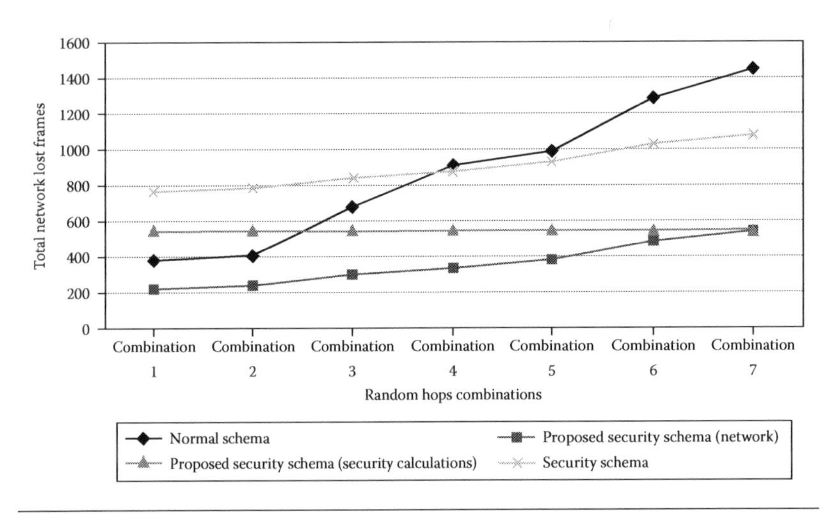

Figure 6.25 The total number of frames lost by the network for four streamed videos with random locations.

3. The network delivery ratio for the proposed security scheme was better when the video camera was deployed farther away from the sink node. The farther the video cameras were deployed from the sink node, the better the delivery ratio.
4. The delivery ratio for the proposed security scheme was always above 65%, which was acceptable for video streaming unless the worst-case scenario occurred.
5. The proposed security scheme fits the light traffic on WMSNs, whether one, two, three, or four video cameras were used.
6. Deciding how many sensor cameras to deploy and where to deploy them would be a network engineer design parameter.

6.4.6.3 Simulation Results for Long Multimedia Streaming The proposed security scheme proved its effectiveness versus the normal scheme for short multimedia streaming. This section will describe the behavior of the proposed security scheme in the simulation when long multimedia content was streamed through the network. The duration of the streamed video here was 120 s, longer than the short video by 100 s.

6.4.6.4 Long Multimedia Streaming with Four Video Cameras Deployed in Different Locations In this scenario, there were four video cameras at different locations in the topology of the network; the video cameras were at a distance of one hop, two hops, three hops, and four hops from the sink. Figure 6.26 represents the energy consumption.

The time at which streaming of the four videos was begun was a random time between 0 and 480. Figure 6.26 shows that the behavior of the energy consumption metric for the proposed security scheme when streaming long multimedia content was the same as when streaming short multimedia content.

Note that, when four videos were streamed, the margin between the proposed security scheme and the normal one became larger. This was a good indicator that the proposed scheme performed well in heavy traffic streaming according to the energy metric.

The good performance of the energy metric of the proposed scheme was because of the proposed scheduler introduced in this work. Figure 6.27 displays the network delivery ratio for four streamed videos with different locations.

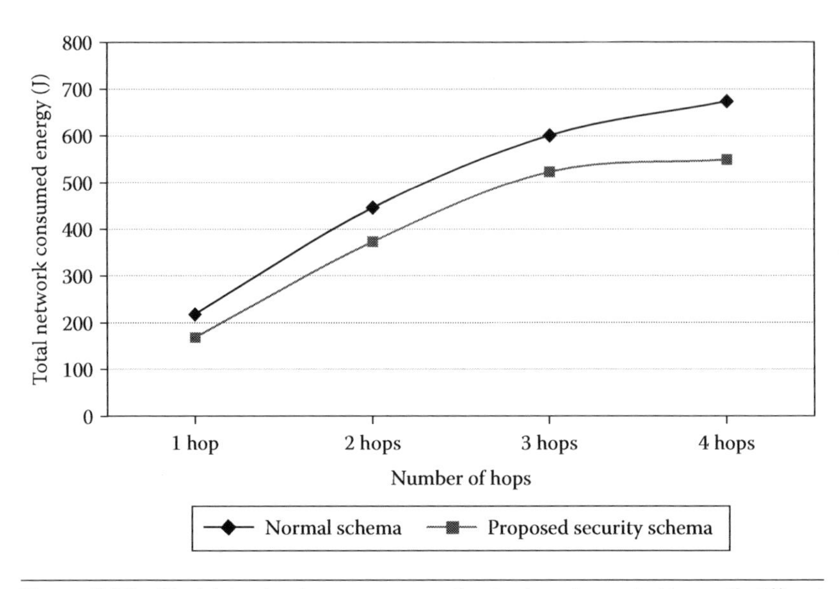

Figure 6.26 The total network energy consumption for four streamed videos with different locations in long multimedia streaming.

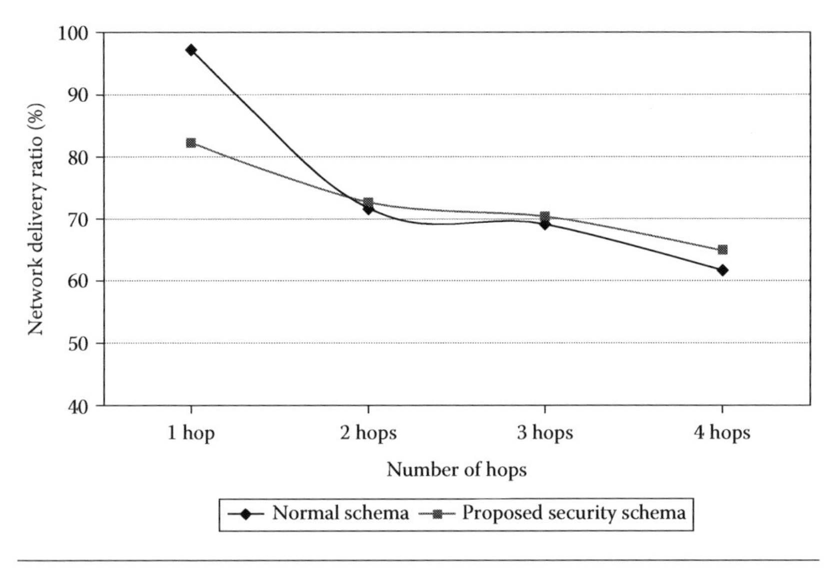

Figure 6.27 The total network delivery ratio for four streamed videos with different locations in long multimedia streaming.

As seen in Figure 6.27, when four video cameras were located one hop away from the sink, the proposed security scheme achieved a delivery ratio of 81%, whereas the normal one achieved 98%. The lower delivery ratio was caused by the security calculations and the proposed scheduler. When the four video cameras were away from the sink by two, three, and four hops, the proposed security scheme achieved a better delivery ratio than the normal one. This better ratio was due to the proposed scheduler.

It was surprising that the proposed security scheme achieved a better delivery ratio when streaming long multimedia content than short content. An investigation of this phenomenon revealed that the reason was the random start time of the video. The random start time range was between 0 and 480 for the long videos. In short multimedia streaming, when the random value for the start time was between 0 and 80, at least three videos would margin with each other, while in long multimedia streaming at least two videos would margin together. This factor made the network more relaxed and enabled a higher delivery ratio. Figure 6.28 shows the total number of lost frames for both the normal scheme and proposed security scheme.

Figure 6.28 demonstrates that the proposed security scheme's lost frames were due to network congestion and the security calculations. The frames lost due to the security calculations were always between 18% and 20% of the original video to be streamed. If the network had more resources, the proposed security scheme would only have lost frames due to the security calculations.

Note that the above scenario was the worst case for the entire simulation done for this study. There was no need to simulate random locations of video cameras, as the worst case achieved 65% delivery ratio.

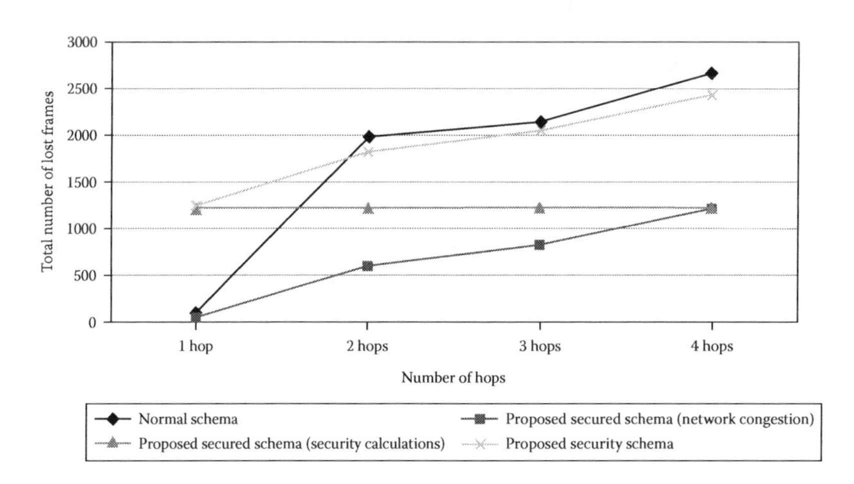

Figure 6.28 The total network of lost frames for four streamed videos with different locations in long multimedia streaming.

6.4.7 PSNR Results for Multimedia Streaming

The PSNR metric had a nonlinear relationship with the subjective video quality. Therefore, it was important to use a utility function that describes subjective video quality as a function of the PSNR. The excellent range of values for optimal video quality is from 30 to 50 dB, while an acceptable range in wireless transmission settled between 20 and 25 dB.

The PSNR metric had to be calculated for every video streamed through the network with every simulation scenario. More than 110 PSNR graph could have been constructed, but the author decided to calculate the PSNR when the worst delivery ratio occurred in the network. The worst delivery ratio was 62% and occurred when four videos were streamed, each four hops away from the sink.

To calculate the PSNR value, both the original video and the resulting streamed video needed to be in YUV format. The researchers of this work used the FFmpeg tool [59] to convert both videos. Figure 6.29 illustrates the PSNR values of the streamed video when the worst-case scenario happened in the network.

The PSNR mean value was 20.43 dB. This value was quite low but acceptable, because the acceptable region for the PSNR value was between 20 and 25 dB for wireless transmission. The PSNR value indicated that the proposed security scheme for multimedia streaming

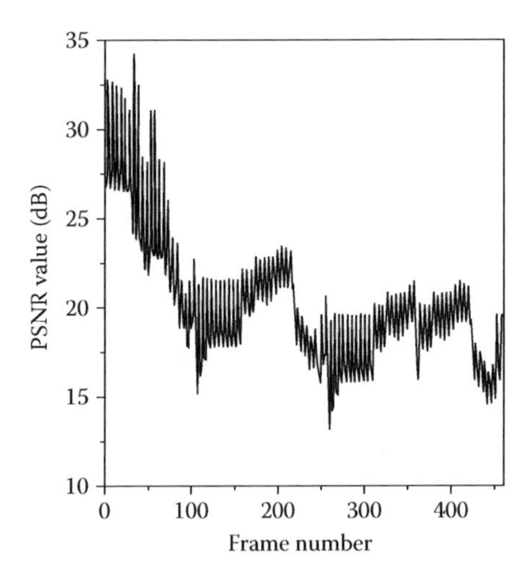

Figure 6.29 The peak signal-to-noise ratio between the original video file and the received video file in the worst-case scenario.

achieved not only lower energy consumption but also an acceptable delivery and PSNR value while proposing a strong security scheme—in other words, four times stronger than any proposed security scheme.

6.4.8 Results Discussion

We have demonstrated that the proposed security scheme for multimedia streaming was more effective than the normal scheme. The effectiveness of the proposed security scheme was measured and tested against the normal scheme according to four metrics—energy consumption, network delivery ratio, frame loss, and PSNR.

The proposed security scheme for multimedia streaming has advantages and disadvantages. The forthcoming points will tackle the advantages of the proposed security scheme for multimedia streaming:

- The proposed security scheme added security features to WMSNs and was four times stronger than any other proposed security scheme.
- Although the proposed security scheme added security features to WMSNs, it achieved lower energy consumption due to the proposed scheduler.
- It achieved an acceptable delivery ratio when video cameras were deployed farther away from the sink node.
- In the worst-case scenario, although the proposed security achieved a delivery ratio of 62%, the PSNR illustrated that the subjective quality of the video was acceptable at 20.43 dB.
- The proposed security scheme was biased to better performance when streaming long content rather than short and achieved better results in delivery ratio.
- The farther the video cameras were deployed, the better the delivery ratio that the proposed security scheme achieved.
- Deciding how many sensor cameras to deploy and where to deploy them should be a network engineer design parameter.

The proposed security scheme for multimedia streaming also had the following disadvantages:

- The proposed security scheme achieved a lower delivery ratio than the normal scheme when the network had only one video camera deployed in the network at all times.

- Moreover, it had a lower delivery ratio than the normal scheme when video cameras were deployed only one hop away from the sink node.

6.4.9 Evaluation of the Proposed Security Operations

Any proposed security scheme will have some trade-off to offer security features. We have already discussed the cost paid by the network in terms of energy consumption, delivery ratio, frame loss, and PSNR.

This section will detail the cost of the security processes. Although NS2 included many features and modules, it did not include any modules for node process calculations in terms of time or energy.

The proposed security scheme based its security features on two processes: the process of encryption and decryption using the AES encryption technique and the process of authentication and deauthentication by the CMAC mechanism.

Refs. [49–51] maintain that both the AES encryption technique and CMAC mechanism are applicable to WSN and WMSN motes. The authors measured the energy consumption and RAM and ROM size for both algorithms; the results are as shown in Tables 6.6 and 6.7.

Table 6.6 Energy Consumed by AES and CMAC Algorithms

ALGORITHM	ENERGY (mJ/byte)
AES encryption	1.62
AES decryption	2.49
CMAC process	1.4

Source: Wang, Y. et al., *IEEE Communications Surveys and Tutorials*, 8, 2–23, © 2006 IEEE.

Table 6.7 Size of Reserved RAM and ROM by AES and CMAC Algorithms

ALGORITHM	RAM (KB)	ROM (KB)
AES	2	10
CMAC	1	6

Source: Wang, Y. et al., *IEEE Communications Surveys and Tutorials*, 8, 2–23, © 2006, IEEE.

The energy consumed by encryption, decryption, and authentication was low compared to that consumed by the transmission and reception functions; data transmission accounts for 70% of the total energy consumption of a WSN [51].

6.4.10 Performance Comparison with Other Related Works

Security in WMSNs is a new field of study in the research community. To date, there have only been a few security frameworks presented for WSNs. The most well-known frameworks for WSN security are TinySec, SenSec (http://ieeexplore.ieee.org/xpl/login.jsp?tp=&arnumber=5961521), MiniSec, TinyECC, and ContikiSec. An overview of these frameworks was given in Section 6.2.6.

Table 6.8 provides references for these frameworks and compares them with the proposed security scheme for multimedia streaming.

6.5 Conclusions

At the beginning of this study, an initial study was carried out to determine the key challenges facing WMSN security. The research objectives became obvious after the initial study, and the main objective was to propose a security scheme that would fit the nature of streaming multimedia in WMSNs. There are a number of frameworks that have been implemented for WSNs that may fit WMSNs, but due to time and budget limitations, the current research was restricted to working on WSN simulators.

The selection criterion for a WSN simulator was provision of security features. Various simulation tools were matched including OPNET, QualNet, GloMoSim, and NS2, none of which supported security features. NS2 was determined to be the most appropriate simulation tool based on its flexibility, in addition to the fact that it was publically available free software. It is still regarded by the research community as the most credible network simulator.

After NS2 was selected as the main simulator for this research, several trials were executed and numerous modifications were made to the NS2 infrastructure to adapt it to the security features of the

Table 6.8 Comparison between the Proposed Security Scheme and Other Security Frameworks

FRAMEWORK NAME	YEAR	IMPLEMENTED/ SIMULATED	SECURITY PROPERTIES	ALGORITHMS	LAYER	OVERHEAD	KEY LENGTH	SUPPORT MULTIMEDIA
TinySec [33]	2004	Implemented (NesC)	Access control, integrity, confidentiality, replay protection	Skipjack CBC-CS mode	Link layer	8 bytes/packet	80	—
SenSec [52]	2005	Implemented (NesC)	Access control, integrity, confidentiality, key management	Skipjack-X CBC-CS mode	Link layer	5 bytes/packet	80	—
MiniSec [34]	2007	Implemented (NesC)	Predeployed symmetric keys, confidentiality, replay protection, authentication	Skipjack OCB mode	Network layer	3 bytes/packet	80	—
TinyECC [35]	2007	Implemented (NesC)	Key Exchange, Public key encryption, Digital signature	ECC SECG-160	Link Layer	Keys overheads	256	—
ContikiSec [36]	2009	Implemented (C)	Authentication, Integrity & Confidentiality	AES CBC-CS mode	Network Layer	4 bytes/packet	128	—
Proposed Security Scheme	2013	Simulated (NS2 – C)	Authentication, Integrity, Confidentiality & freshness	AES CMAC Authentication	Transport and Application layer	4 bytes/packet + 2 bits/ packet	256 – 4 times stronger than TinyECC	Yes

proposed security scheme. More than 200 hours of development and simulation were put into implementing the security features in NS2 for multimedia streaming.

At the end of the development and implementation phase, comparisons between the proposed security scheme and the normal one were drawn according to the network delivery ratio and energy consumption, which were the main interests of this study. Moreover, a comparison was made between other frameworks for security in WSNs and the proposed security scheme.

The most significant contributions of this study are as follows:

- It managed to add security libraries to NS2, which did not support security features before. Appendix 6A illustrates the steps that were taken to integrate security features into NS2.
- It attempted to propose a security scheme for WMSNs for multimedia streaming.
- The proposed secured scheme for multimedia streaming achieved a significant savings in network energy consumption, consuming 15%–20% less energy consumption than the normal scheme despite the addition of security features. This significant savings was due to the proposed scheduler for WMSNs, which predicts video frames that cannot handle the process of encryption, decryption, and authentication and discards them before sending them through the network.
- The suggested security scheme for multimedia streaming outperforms the normal scheme in delivery ratio when the sensor cameras are farther away from the sink node. The farther away the video cameras are deployed, the better the delivery ratio provided by the proposed security scheme. Moreover, the proposed security scheme achieved acceptable subjective video quality (PSNR value), even though the main goal of the research was energy consumption.
- The proposed security scheme was four times stronger than other security frameworks: every packet contains a PW field with four values, which makes the proposed security scheme computationally secure, but is considered as overhead by 2 bits/packet.

6.6 Implications for Future Works

There are still several research challenges that need to be addressed before use of the proposed security scheme in real WMSN deployment. The following list will offer some implications for future research related to the study presented in this chapter:

- Simulating the proposed security scheme with other application and transport layer protocols, for example, TCP, FTP, and HTTP
- Building a computational processing model for NS2 to calculate the consumed energy for the code executed inside WMSN nodes
- Concerning the proposed work in multimedia streaming, extending the video codec used in this research to such video codecs as MPEG-4, H.261, H.264, and so on
- Comparing the AES encryption algorithm with video encryption algorithms by means of using NS2
- Implementing the proposed security scheme in WMSN operating systems, such as TinyOS or Contiki, and comparing it with the current security frameworks implemented on these operating systems

Finally, it is hoped that the research presented in this chapter will help other researchers to simulate their security ideas and mechanisms based on the suggested development and the integration of security libraries into NS2.

Appendix 6A: Proposed Security Scheme Implementation

This appendix will provide a detailed description of the implementation of the proposed secured scheme using NS2. It will describe how to install NS2 and how to implement multimedia streaming into it within the proposed secured scheme.

6A.1 NS2

NS2 is a simulation tool primarily targeted for networking research and educational use. The simulator was derived from the old network simulator Real and Large (REAL) in 1989, which was developed with the goal of studying flow and congestion-control schemes in packet-switched data networks [44].

In 1995, the first generation of NS was completed through the VINT project with the hope of becoming a common simulator with advanced features to change the then-prominent protocol engineering practices [44]. The simulator continued to evolve and the second generation, NS2, was first released in 1996 based on NS1.

NS2 is extensively used by the networking research community. It provides substantial support for simulation of TCP, UDP, routing, multicast protocols over wired and wireless (local, satellite, sensor) networks, and so forth. [44]. The simulator is event-driven. It consists of C++ core methods and uses Tool Command Language (Tcl) and Object Tcl shell as interface allowing the input file (simulation script) to describe the model to simulate. Users can define arbitrary network topologies composed of nodes, routers, links, and shared media [44].

6A.1.1 NS2-Supported Platforms

NS2 is supported by many platforms; it runs over the Linux, FreeBSD, and Solaris operating systems. It is also supported in Windows 98/2003/XP/Vista/7 operating systems, but it requires the Cygwin platform [53]. The Cygwin platform is a collection of tools that provide an environment with a Linux look and feel for the Windows operating system. Windows 2003 with the Cygwin platform was the preferred choice for the present study, due to the researchers' experience with Windows 2003.

The first step in the implementation phase was the installation of the Cygwin platform. This step was achieved by following the tutorials in Ref. [54]. This reference gives excellent instructions for how to install the Cygwin platform on a Windows operating system.

6A.1.2 NS2 Installation

The second step is installing the NS2 simulator over the Cygwin platform, which was installed on a Windows 2003 operating system. The researchers installed NS version 2.29. It was the latest version of NS2 at the time. Ref. [55] describes how to install NS2 over the Cygwin platform.

6A.1.3 NS2 Layered Structure Modifications

The implementation of the proposed security scheme in this research for both message transmission and multimedia streaming is developed across the layers of NS2. Figure 6A.1 represents the implementation modifications that have been made to adapt the proposed secured scheme.

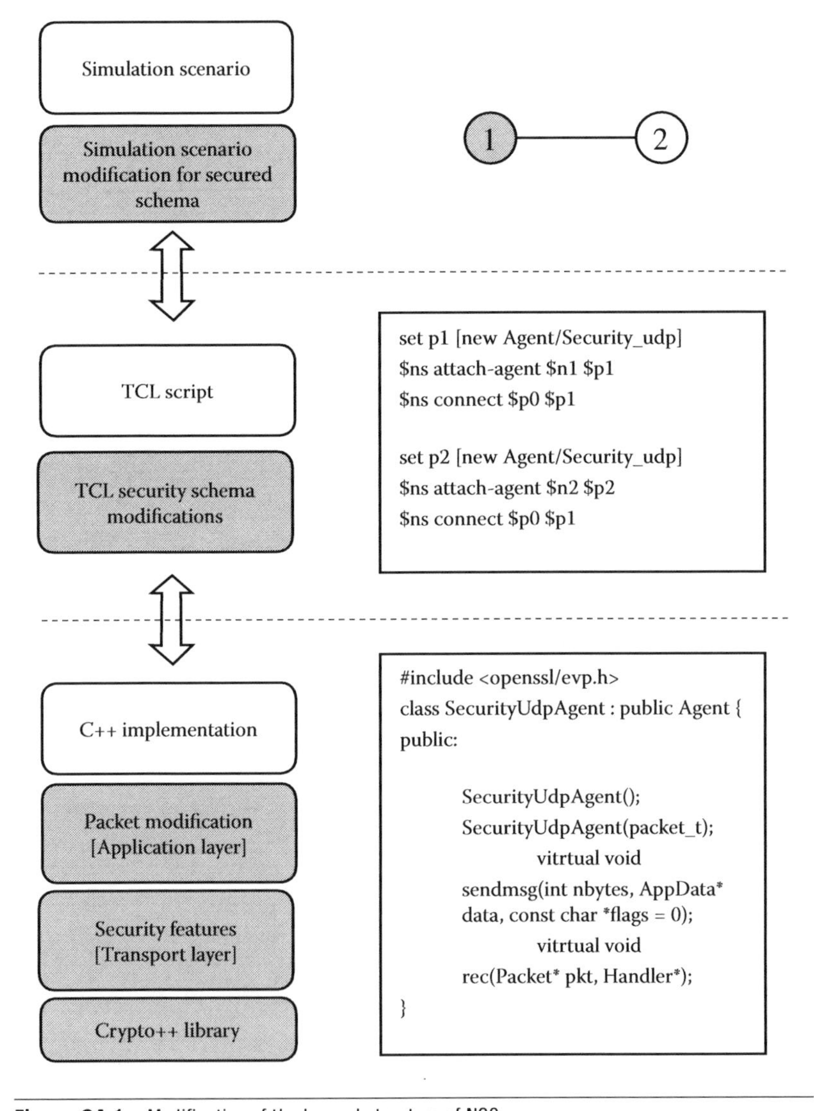

Figure 6A.1 Modification of the layered structure of NS2.

Implementing the modifications of the proposed security scheme detailed in Section 6.3.2 can be summarized by the following steps:

1. Add the Crypto++ library [56] to NS2 libraries and structure (C++ implementation layer).
2. Modify the packet format to add MAC-generated code (C++ implementation layer).
3. Modify the UDP to adapt the security features mentioned in the proposed security scheme, namely, CMAC generation, AES encryption, and C++ implementation layer.
4. Add the proposed security scheme to the Tcl script layer to deal with it in the simulation scenario layer (Tcl layer).
5. Call the proposed security scheme from the simulation scenario layer and initialize simulation parameters.

6A.1.3.1 C++ Implementation Layer Modification The modification of the C++ implementation layer consists of two steps. The first step is adding the Crypto++ library to the NS2 C++ implementation layer, whereas the second step involves adding CMAC code to the WSN packet format and encrypting the payload data with the AES encryption technique.

Crypto++ is a free open-source C++ class library of general purpose cryptographic algorithms and schemes. It provides a C++ application programming interface (API) for cryptographic functionality, such as encryption and decryption, hashing, key agreement schemes, random number generation, and others [56]. Adding the Crypto++ library to the NS2 C++ implementation layer was achieved after several trials and took more than 50 hours of research time. This length of time was a result of the fact that NS2 will not accept a new library unless it is integrated into its structure without any errors. The following points were applied to implement this step:

1. Download the Crypto++ library from Ref. [56]
2. Set the downloaded library in the Cygwin library directory
3. Select Install File in the NS2 simulator and add the Crypto++ library to the NS2 library
4. Install NS2 again with the modified NS2 install file
5. Select Make File in NS2 and add the line {-lm -lm -lcrypto} to the library code line
6. Build the whole NS2 simulator and run the make file

The second step, adding CMAC code to the WSN packet format and encrypting payload data, required numerous attempts. The author investigated NS2 books to study its structures. The work done in the present study concentrated on Ref. [54], which provides a detailed account of how to add new protocols into the structure of NS2 without facing errors. The following points describe how to implement this second step:

1. For WMSN packet modification, create a duplicate copy of the files rtp.h and rtp.cc in the Apps folder in the NS2 directory.
2. Name these copies "secured_rtp.h" and "secured_rtp.cc", respectively.
3. Open secured_rtp.h and rename all named RTP names to "secured_rtp".
4. Open secured_rtp.cc and rename all named RTP names to "secured_rtp".
5. Add a char variable of 4 bytes of cmac_code in the hdr_secured_rtp structure.
6. Add a 2-bit variable of PW in the hdr_secured_rtp structure.
7. The CMAC code generation will be added to the received and sent functions of the proposed secured scheme.
8. For the proposed secured scheme, create a duplicate copy of udp.h and udp.cc in the Apps folder in the NS2 directory.
9. Name these copies "secured_udp.h" and "secured_udp.cc", respectively.
10. Open secured_udp.h and make the following changes:
    a. Include the security library by the following line: {#include <openssl/evp.h>}.
    b. Define the block of AES by the following line: {#define AES_BLOCK_SIZE 256}.
    c. Add the following functions of AES encryption and decryption to the class of secured_upd.

```
int aes_initiation(unsigned char *key_d, int key_d_len,
  unsigned char *salt, E_C_CT*e_ct, E_C_CT*d_ct);
unsigned char *aes_encryption(E_C_CT *e, unsigned char
  *pt, int *len);
unsigned char *aes_decryption(E_C_CT *e, unsigned char
  *ct, int *len);
```

11. Open secured_udp.cc and make the following changes:
    a. Initiate AES using the aes_initiation function in the constructor of secured_upd class.
    b. Modify the send function of secured_udp by calling the aes_encryption function, then call the function CMAC and pass the encrypted payload and WSN fields to it. The generated code will be appended the secured_rtp.
    c. Modify the receive function of secured_udp by calling the function CMAC and pass the encrypted payload and WSN fields to it. If the generated code is equal to the CMAC code in the header, then call the aes_decryption function or report attack.

6A.1.3.2 Tcl Layer Modification The modification in the second layer (the Tcl layer) is related to the first layer (the C++ implementation layer). The modifications to the second layer may be implemented as follows:

1. Modify the file {tcl/lib/ns-default.tcl} inside the NS2 directory and add the following lines:

```
Agent/Security_udp instproc done {} { }
Agent/Security_udp instproc process_data {from } { }
```

2. Modify the file {tcl/lib/ns-packet.tcl} inside the NS2 directory and add the line in bold:

```
foreach prot {
AODV
# others:
Security_udp # Proposed Security Schema
}
```

3. Modify the file located in {Makefile} inside the NS2 directory and add the following line:

```
apps/Security_udp.o \
```

The modification in the file {tcl/lib/ns-default.tcl} is considered a mirror of the modification made in the first layer (the C++

implementation layer). The other lines are added to define the packet size and process data in the proposed security scheme in the form of text messages. The modification in the file {tcl/lib/ns-packet.tcl} is made to define the new secured packet format in the file, which contains all packet formats on NS2 protocols.

The modification in the file {Makefile} is the link between the first layer (the C++ implementation layer) and the second layer (the Tcl layer), which notifies the C++ implementation layer that a new protocol has been added to NS2 protocols.

After the above modifications are carried out, NS2 must be recompiled. The recompile of NS2 is needed to execute the changes that have been developed in the NS2 structure.

6A.1.3.3 Simulation Scenario Layer Modification After applying the modification to the first layer (the C++ implementation layer) and the second layer (the Tcl layer), the implementation of the proposed security scheme is complete. The last step is building a simulation scenario of the proposed security scheme in the Simulation Scenario layer. In order to call the new proposed security scheme, you can add the following lines to any simulation scenario. The following lines describe the calling of the proposed security scheme; for example, in this case WMSN node n1 is attached to WMSN node n0 {sink node}.

```
#Create two secured agents of the proposed secured
 schema and attach them to the nodes n0 and n1
set p0 [new Agent/Security_udp]
$ns attach-agent $n0 $p0
#-------------------------------------------------------
set p1 [new Agent/Security_udp]
$ns attach-agent $n1 $p1
$ns connect $p0 $p1
```

6A.2 Multimedia Streaming Modification

To study the problems mentioned above in greater detail and propose a solution for them, the author implemented the steps proposed by Chih-Heng Ke [57]. Chih-Heng Ke proposed a tool, called "EvalVid," that can be integrated with NS2 to stream multimedia in real time over a WMSN. Figure 6A.2 represents the EvalVid tool components.

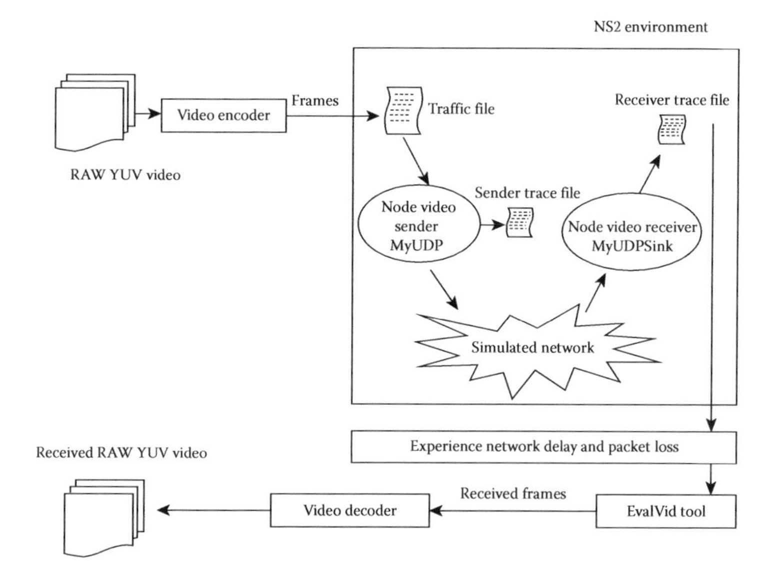

Figure 6A.2 Design diagram of the EvalVid tool. (From Ke, Chih-Heng et al. *J. Inf. Sci. Eng.* 24.2 (2008): 425–440.)

MyUDP is an extension of the UDP agent for the node video sender. This new agent allows users to specify the output file name of the sender trace file, and it records the timestamp of each transmitted packet, the packet ID, and the packet payload size. The task of the MyUDP agent corresponds to the task that tools such as RTP-dump do in a real network environment [57].

MyUDPSink is the receiving agent of the receiver node for the fragmented video frame packets sent by MyUDP. This agent also records the timestamp, packet ID, and payload size of each received packet in the user-specified file [57].

After installing the EvalVid tool in NS2 and making sure that it was working without errors, the author began to investigate the code and to modify it to add security features as mentioned in the proposed security scheme in Section 6.3.2.

6A.2.1 *Secured Version for EvalVid Tool*

The secured design of the EvalVid tool is presented in Figure 6A.3. The main changes are 1) the proposed mechanism was added to the NS2 scheduler and 2) the security feature of the proposed security scheme was added to both the video sender and receiver.

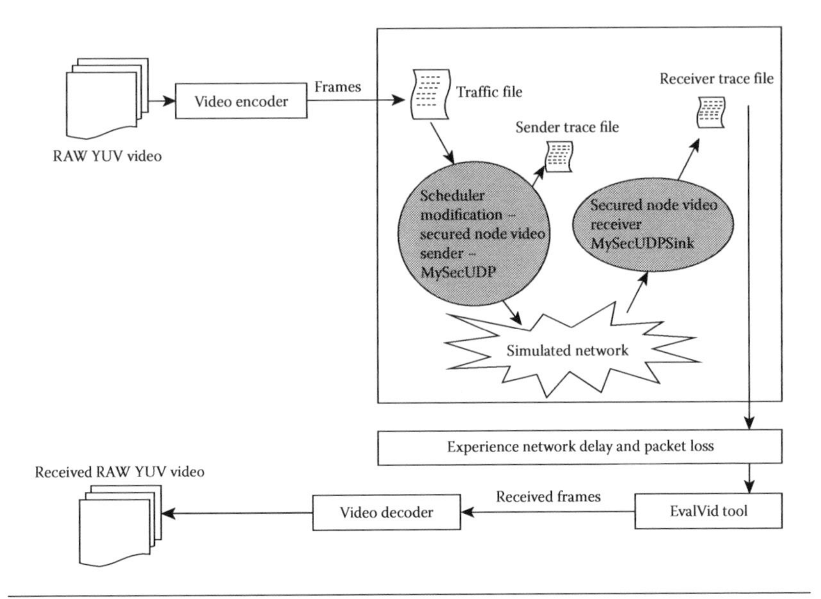

Figure 6A.3 Design diagram of secured EvalVid.

Figure 6A.4 Screenshot of both videos used in simulations. (From National Science Foundation, Arizona State University, 2000.)

6A.3 Experiment Data Preparations

The researchers of this work used standard video sequences that are commonly used for multimedia experiments. The video sequences can be found at the following website: http://trace.eas.asu.edu/yuv/index.html [58].

The standard video used in short multimedia streaming can be found at the above-mentioned site with the name "Container." The standard video used in long multimedia streaming can also be reached on the same website with the name "Bridge (Close)." A screenshot for both videos is displayed in Figure 6A.4.

Both videos are presented in raw YUV sequence format. The size of the short multimedia streaming video (Container) is 45 megabytes, whereas the long multimedia streaming video (Bridge [Close])

1364566535.613739 RTP len=1328 from=193.227.14.69:1728 v=2 p=0 x=0 cc=0 m=0 pt=33 (MP2T,0,90000) seq=17574 ts=764884950
ssrc=0x6062
data=4700441830087810a0009804c3d8426704116010703e9049fd900d8a0244f862eeb003f01593a0007a50c01288ae0411600b80250d049fd4cdc
01a006408bf64d008fc005881d005ee9c031cc01381c814027703f6904114010022fe10c2d31c00f085ba920549f303081d30980000010713e521042
2804a3cc48fab014013482481a6ee003d040b1006201a00c434988210064a4f3c0a61f5a0835000ac1141803004907d2d17e6f13099c113e90149204
b8040189044fac0112481470044197a5041aa005608a0a01802484620cc9040ac005690440400d038990a1f17e601046807430349a58150c28a2b3a
529417941f7ad040d7040d42804f8004c00f10016006451259490189604cb01e0ed6e006e081a400cc01993124202a02748184a5c0c2158408badeae
f4e50eba48403be4c2186135868182d09183dbdf37e08168008007690074031417c06e42e05c0744d412908ecce064e55f4e1b177a9242259b3e5fbf
98940819800c099c04e3001b809834033010ac4700441a9a03b4b801f00e8ac480c267602e330042bbaa005608188018600700822800b121a1a4b70
0ac04e9412897d20391802b235ee33dc704033005c9004a00ac040030f8050349658e1852e8298999440b6e0302596352bbf0c0805600a0004800c0
04c3402726e01b96028dcf38fb9a4a1860250222c65f41082199959072b5a8b00cc07485a58127f711a961a6d992156e91a4dc089f480152a24c00b4
306e042fa2dc09e7fa2083c80191c84164d141986ca5006a0310d01c4700441b0060595879696f0103f43fba09a422193101890094c199c3eb096430
090341240f55621a460218296362016934090c81904bf6ee054706856a380c4090693aac9658fab2b00f644a912eef03203b024980d01df313509260
0f665231a1704a9819bb98863933701b8e485d512f80484fa604704c170062097768217b8e0b4ca4a013869e4eaa74a4908d40321b80fcc95036e200
749412406fc412a54908998117f002360595879696f0103f43fba09a422193101890094c19960595879696f0103f43fba0605958

Figure 6A.5 An example of a frame to be sent by a node in multimedia streaming simulations.

is 297 megabytes. The YUV format cannot be used in WMSNs because the size of frames is very large. To overcome this problem, the researchers applied a multimedia encoder to encode the video files into the standard encoding format used in RTP protocol. The standard RTP encoding is moving picture experts group (MPEG). In the current work, the FFmpeg [59] encoder was used to convert video files to MPEG format. The FFmpeg encoder is a cross-platform solution for recording, converting, and streaming audio and video files. The following commands are used to convert both videos using the FFmpeg encoder.

```
ffmpeg -s cif -i container_cif.yuv -qscale 7 -r 40 container.mpeg
ffmpeg -s cif -i bridge-far_cif.yuv -qscale 7 -r 40 bridge-far.mpeg
```

The –qscale was equal to 7 to reduce the quality of the video from 10 to 7. The –r was set equal to 40, as 40 is the number of frames per second. The previous command was achieved after several trials, as FFmpeg has about 70 parameters to be used while converting video files.

The newly produced sizes for both videos were 1.4 megabytes for the file Container and 5.9 megabytes for the file Bridge (Close). After converting files into MPEG format, these files were used in the NS2 simulation. One of the frames is shown in Figure 6A.5.

References

1. Akyildiz, I.F., Vuran, M.C. *Wireless sensor networks*, First Edition, John Wiley and Sons, New Jersey, USA, 2010.
2. Sohraby, K., Minoli, D., Znati, T. *Wireless sensor networks: Technology, protocols, and applications*, First Edition, John Wiley and Sons, New Jersey, USA, 2007.
3. Akyildiz, I.F., Su, W., Sankarasubramaniam, Y., Cayirci, E. A survey on sensor networks, *IEEE Communications Magazine*, 40(8), 102–114, 2002.
4. Gupta, A.K., Sadawart, H., Verma, A.K. Performance analysis of AODV, DSR & TORA routing protocols, *International Journal of Engineering and Technology*, 2(2), 226–231, 2010.
5. Prasanna, S., Rao, S. An overview of wireless sensor networks applications and security, *International Journal of Soft Computing and Engineering*, 2(2), 2231–2307, 2012.
6. Prabh, K.S. Real-time wireless sensor networks, *PhD Thesis*, Faculty of the School of Engineering and Applied Science, University of Virginia, May 2007.

7. Lu, C., Blum, B.M., Abdelzaher, T.F., Stankovic, J.A., He, T. RAP: A real-time communication architecture for large-scale wireless sensor networks. In: *IEEE Real-Time and Embedded Technology and Application Symposium*, IEEE, San Jose, CA, USA, pp. 55–67, September 2002.
8. He, T., Stankovic, J.A., Lu, C., Abdelzaher, T.F. SPEED: A real-time routing protocol for sensor networks, *Technical Report CS-2002-09*, Virginia Polytechnic Institute and State University, Virginia, USA, March 2002.
9. Sinha, A., Chandrakasan, A. Dynamic power management in wireless sensor networks, *IEEE Design and Test of Computers*, 18(2), 62–74, 2001.
10. Mottola, L., Picco, G. Programming wireless sensor networks: fundamental concepts and state of the art, *ACM Computing Surveys*, 43(3), 1–51, 2011.
11. Abdelzaher, T., Blum, B., Cao, Q., Chen, Y., Evans, D., George, J., George, S. et al. EnviroTrack: Towards an environmental computing paradigm for distributed sensor networks. In: *24th International Conference on Distributed Computing Systems*, Department of Computer science and Engineering, Ohio State University, Tokyo, Japan, pp. 582–589, 2004.
12. Mohammadi, S.H., Jadidoleslamy, H. A comparison of link layer attacks on wireless sensor networks, *International Journal on Applications of Graph Theory in Wireless Ad Hoc Networks and Sensor Networks*, 3(1), 69–84, 2011.
13. Stallings, W. *Cryptography and network security*, Fourth Edition, Prentice Hall, New Jersey, USA, 2006.
14. Agrawal, M. A comparative survey on symmetric key encryption techniques, *International Journal on Computer Science and Engineering*, 4(5), 877–882, 2012.
15. Koblitz, N., Menezes, A.J. A survey of public-key cryptosystems, *SIAM Review*, 46(4), 599–634, 2004.
16. Black, J.R. Message authentication codes, *PhD Thesis*, Computer Science Department, University Of California, California, USA, 2000.
17. Bakhtiari, S., Safavi-Naini, R., Pieprzyk, J. Cryptographic hash functions: A survey, *Technical Report TR 95-09*, Computer Science Department, University of Wollongong, New South Wales, USA, 1995.
18. Sandström, H. A survey of the denial of service problem, *Master Thesis*, Computer Science and Electrical Engineering Department, Lulea University of Technology, Kiruna, Sweden, 2001.
19. Cayirci, E., Rong, C. *Security in wireless ad hoc and sensor networks*, First Edition, John Wiley and Sons, New Jersey, USA, 2009.
20. Kavitha, T., Sridharan, D. Security vulnerabilities in wireless sensor networks: A survey, *Journal of Information Assurance and Security*, 5(1), 31–44, 2009.
21. Deng, J., Han, R., Mishra, S. Defending against path-based DoS attacks in wireless sensor networks. In: *3rd ACM Conference on Security of Ad Hoc and Sensor Networks*, Association for Computing Machinery, New York, NY, USA, pp. 89–96, November 2005.
22. Singh, V.P., Jain, S., Singhai, J. Hello flood attack and its countermeasures in wireless sensor networks, *International Journal of Computer Science Issues*, 7(11), 23–27, 2010.

23. Karlof, C., Wagner, D. Secure routing in wireless sensor networks: Attacks and countermeasures, *Ad Hoc Networks*, 1(3), 293–315, 2003.
24. Huang, J.H., Buckingham, J., Han, R. A level key infrastructure for secure and efficient group communication in wireless sensor network. In: *1st International Conference on Security and Privacy for Emerging Areas in Communications Networks*, IEEE, Athens, Greece, pp. 249–260, September 2005.
25. Brownfield, M., Yatharth, G., Nathaniel D. Wireless sensor network denial of sleep attack. In: *6th Annual IEEE Systems, Man and Cybernetics (SMC) Information Assurance Workshop*, IEEE, New York, NY, USA, pp. 356–364, June 2005.
26. Rgheff, A., Ali, M. Fundamentals of spread-spectrum techniques, 2007. Last accessed May 2013. http://v5.books.elsevier.com/bookscat/samples/9780750652520/9780750652520.pdf
27. Islam, M.R. Error correction codes in wireless sensor network: An energy aware approach, *International Journal of Computer and Information Engineering*, 4(1), 59–64, 2010.
28. Li, M., Prabhakaran, B. MAC layer admission control and priority re-allocation for handling QoS guarantees in non-cooperative wireless LANs, *Mobile Networks and Applications*, 10(6), 947–959, 2005.
29. Cho, J., Lee, J., Kwon, T., Choi, Y. Directional antenna at sink (DAAS) to prolong network lifetime in wireless sensor networks. In: *12th European Wireless Conference for Enabling Technologies on Wireless Multimedia Communications*, IEEE, Athens, Greece, pp. 1–5, April 2006.
30. Hue, J. Deluge 2.0 - TinyOS network programming, 2005. Last accessed May 2013. http://www.cs.berkeley.edu/~jwhui/deluge/deluge-manual.pdf.
31. Hyun, S., Ning, P., Liu, A., Du, W. Seluge: Secure and DoS-resistant code dissemination in wireless sensor networks. In: *7th International Conference on Information Processing in Sensor Networks*, New York, NY, USA, pp. 445–456, April 2008.
32. Pathan, A.K. *Security of self-organizing networks: MANET, WSN, WMN, VANET.* First Edition, Auerbach Publications, 2010.
33. Karlof, C., Sastry, N., Wagner, D. TinySec: A link layer security architecture for wireless sensor networks. In: *2nd International Conference on Embedded Networked Sensor Systems*, Association for Computing Machinery, New York, NY, USA, pp. 162–175, November 2004.
34. Luk, M., Mezzour, G., Perrig, A., Gligor, V. MiniSec: A secure sensor network communication architecture. In: *6th International Conference on Information Processing in Sensor Networks*, Association for Computing Machinery, New York, MA, USA, pp. 479–488, April 2007.
35. Liu, A., Ning, P. TinyECC: A configurable library for elliptic curve cryptography in wireless sensor networks. *Technical Report TR-2007-36*, Department of Computer Science, North Carolina State University, North Carolina, USA, November 2007.

36. Casado, L., Tsigas, P. ContikiSec: A secure network layer for wireless sensor networks under the Contiki operating system. In: *14th Nordic Conference on Secure IT Systems*, Springer - International Publisher Science, Oslo, Norway, pp. 133–147, October 2009.
37. Black, J., Rogaway, P. CBC MACs for arbitrary-length messages: The three-key constructions, *Journal of Cryptology*, 18(2), pp. 111–132, 2005.
38. Mahmoud, N. Securing real-time video over IP transmission, *Master Thesis*, Department of Information Technology, Cairo University, Giza, Egypt, May 2009.
39. Yasmin, R. An efficient authentication framework for wireless sensor networks, *PhD Thesis*, College of Engineering and Physical Sciences, The University of Birmingham, Birmingham, UK, November 2012.
40. Schulzrinne, H., Casner, S., Frederick, R., Jacobson, V. RFC 1889 – RTP: A transport protocol for real-time applications, *Internet Engineering Task Force*, IETF RFC 1889, January 1996.
41. Fitzek, F., Reisslein, M. A prefetching protocol for continuous media streaming in wireless environments, *IEEE Journal on Selected Areas in Communications*, 19(6), pp. 2015–2028, 2001.
42. Levis, P., Madden, S., Gay, D., Polastre, J., Szewczyk, R., Whitehouse, K., Woo, A. et al. TinyOS: An operating system for sensor networks. In: *Ambient Intelligence*, Springer-Verlag, New York, NY, USA, pp. 115–148, 2004.
43. Eriksson, J. Detailed simulation of heterogeneous wireless sensor networks, Department of Information Technology, Uppsala University, PhD Thesis, Thunbergsvägen, Sweden, April 2009.
44. Kellner, A., Behrends, K., Hogrefe, D. Simulation environments for wireless sensor networks, *Technical Report TR IFI-TB-2010-04*, Institute of Computer Science, Georg-August University, Göttingen, Germany, June 2010.
45. Sobeih, A., Chen, W., Hou, J.C., Kung, L., Li, N., Lim, H., Tyan, H. et al. J-Sim: A simulation and emulation environment for wireless sensor networks. *IEEE Wireless Communications Magazine*, 13(4), 104–19, 2006.
46. Xue, Y., Lee, H.S., Yang, M., Kumarawadu, P., Ghenniwa, H., Shen, W. Performance evaluation of ns-2 simulator for wireless sensor networks. In: *Canadian Conference on Electrical and Computer Engineering*, IEEE, Vancouver, BC, Canada, pp. 1372–1375, April 2007.
47. Khater, J. NS-2 simulation based study of E2E video streaming over ultra-wideband (UWB) wireless mesh networks, *Master Thesis*, Athens Information Technology, Peania, Athens, 2006.
48. Khan, A., Lingfeng, S., Ifeachor, E. Impact of video content on video quality for video over wireless networks. In: *5th International Conference on Autonomic and Autonomous Systems*, Valencia, Spain, pp. 277–282, May 2009.
49. Lee, J., Kapitanova, K., Son, S.H. The price of security in wireless sensor networks, *Computer Networks*, 54(17), pp. 2967–2978, 2010.

50. Wang, Y., Attebur, G., Ramamurthy, B. A survey of security issues in wireless sensor networks, *IEEE Communications Surveys and Tutorials*, 8(2), pp. 2–23, 2006.
51. Bala, S., Secure routing in wireless sensor networks, *Master Thesis*, Computer and Engineering department, Thapar University, Punjab, India, May 2009.
52. Tieyan, L., Hongjun, W., Xinkai, W., Feng, B. SenSec design, sensor network flagship project, *Technical Report TR v1.0*, InfoComm Security Department, Institute for Infocomm Research in Singapore, February 2005.
53. Faylor, C. Cygwin, 1995. Last accessed May 2013. http://www.cygwin.com/
54. Yan, S.F. How to implement protocol in NS2. Last accessed May 2013. http://netlab.cse.yzu.edu.tw/ns2/html/chu-workshop/session2-2.pdf
55. The Vint Project. Network Simulator - ns-2, 1989. Last accessed May 2013. http://www.isi.edu/nsnam/ns/
56. Bider, D. Crypto ++ library, 2007. Last accessed May 2013. http://www.cryptopp.com/
57. Chih-Heng, K., Shieh, C., Hwang, W., Ziviani, A. An Evaluation Framework for More Realistic Simulations of MPEG Video Transmission, *Journal of Information Science and Engineering*, 24(2), pp. 425–440, 2008.
58. National Science Foundation, Arizona State University. YUV Video Sequences Datasets, 2000. Last accessed May 2013. http://trace.eas.asu.edu/yuv/
59. Bellard, F. FFmpeg: Software to record, convert and stream audio and video, 2004. Last accessed May 2013. http://www.ffmpeg.org/

7

Power Management for Wireless Multimedia Sensor Networks

RAGHIED MOHAMMED ATTA

Contents

Wireless multimedia sensor networks (WMSNs) cover several branches of communication, computation with signal processing, and control with embedded computing. This cross-disciplinary research field enables distributed systems of heterogeneous embedded devices that sense, interact, and control the physical environment. There are several factors that mainly influence the design of WMSNs, such as high bandwidth demand, integration with other wireless technologies, and power consumption.

Batteries have been the source of energy for most mobile, embedded, and remote system applications. To prolong the lifetime of batteries, hardware power optimizations have been the focus of a vast amount of research in WMSNs. This research includes dynamic optimization of voltage and clock rate, wake-up procedures to keep power-consuming electronics inactive most of the time, and energy-optimized protocol development for sensor communications.

It is clear that alternative sources of energy to power up WMSNs are required. There is a need to supply energy to prolong the lifetime of a system with self-powered devices that use energy-harvesting techniques. Such energy supply might be achieved by extracting energy from the environment where the sensor itself lies, which offers another important means to prolong the lifetime of sensor devices. There are various forms of energy that can be harvested, such as thermal, mechanical, magnetic, solar, acoustic, wind, and water.

Collecting energy from the background environment might provide an additional source of energy to help prolong the lifetime of the sensor devices. However, it will yield power that is several orders of magnitude lower compared to the power consumption of state-of-the-art multimedia devices.

In this chapter, we will look into various ways of optimizing the power consumption of existing WMSNs. We will also identify various unconventional sources of energy that can be harvested based on the available techniques and compare their advantages.

7.1 Introduction

A WMSN is defined as a network of wireless embedded devices that allow retrieval of audio and video streams, images, and scalar sensor data from the physical environment. More recently, there have been rapid improvements in miniaturization of inexpensive hardware such as CMOS cameras and microphones that are able to universally capture multimedia content from the environment; a single sensor device module has enhanced the development of WMSNs [1,2].

A critical issue for most of the applications for which WMSNs are installed is battery replacement; therefore, battery lifetime would ideally be unlimited [3]. The energy capacity of batteries has doubled roughly every 35 years [4]. To a large extent, this development has accelerated in recent years, due to the needs of portable electronic devices; however, the rate of improvement is still fairly slow compared to Moore's law.

Therefore, power consumption is a fundamental concern in WMSNs, even more than in traditional wireless sensor networks (WSNs). In traditional sensor nodes, energy consumption is dominated by communication functions, while multimedia applications produce large amounts of data, which require high transmission rates with large bandwidth and extensive processing. Thus, both power and bandwidth are even more constrained than for other types of WSNs. In the meantime, the faster processors used in WMSNs tend to require more power to operate. It is necessary to come up with techniques to reduce and manage power consumption in high-speed networks.

Sensing power varies with the nature of applications. In applications such as environmental and habitat monitoring, WMSNs are positioned in remote and inaccessible regions such as mountains, forests, deserts, and rural areas to collect multimedia information for an extended duration. Such applications typically consume more energy than in the case of traditional WSNs. Irregular sensing might consume less power than constant event monitoring; however, the complexity of event detection also plays a crucial role in determining power consumption. For example, for the visual processing component, which focuses on detecting and characterizing events within a node camera's field of view, the more images acquired, the more accurate its decision of whether or not an event worth reporting is needed. However, the more images are taken, the more power will be consumed. Higher ambient noise levels might also cause significant corruption and increase detection complexity.

The collapse of a few nodes in the network because of power loss can cause significant topological changes and might require rerouting of packets and reorganization of the network. Therefore, the following issues are critical: a proper management strategy for residual energy; an optimal choice of energy-aware routing protocols, algorithms, and architectures; adequate topological placement of the sensors to maximize the network lifetime; and an efficient energy-harvesting mechanism from the deployment environment [5]. Consequently, there have been investigations into lowering the energy consumption of wireless networks. Researchers have concentrated on developing low-power techniques at all levels, from designing energy-efficient circuits [6] to adapting central processing unit (CPU) frequencies [7] and enhancing network protocols [8].

Another important aspect addressed in WMSNs is whether it is more efficient to send the full video stream or to perform the processing on board, hence saving bandwidth by transmitting only a higher level representation of the data. To make such decisions, it is necessary to know all the information about the power budget of the nodes involved, the power consumed by different operations involved, and the application performance requirements.

7.2 Energy Characterization

WMSN nodes have four major components:

- A microcontroller (MCU), consisting of the processor itself, memory (RAM and flash), and associated hardware
- A radio system, consisting of wireless communication circuitry and an antenna
- A sensor system, consisting of multimedia sensing devices
- A power supply subsystem, consisting of the battery, a DC–DC converter, and an energy harvester

These different components can be in different states. A set of basic operations can be considered representative of tasks performed by a sensor node. These operations consist of five main task categories, namely idle, processing intensive, storage intensive, communication intensive, and visual sensing [9].

1. Idle: Idle state behavior consumes energy when the node runs basic operating system tasks. In addition to characterizing energy consumption when the system is idle, this task also serves as a reference for all other tasks.
2. Processing intensive: Fast Fourier transform, an industry-standardized CPU-intensive efficient algorithm, is used to characterize processing-intensive tasks.
3. Storage intensive: The storage media is a memory that can store and recall files as input parameters.
4. Communication intensive: Energy consumed by communication-related tasks can be characterized by transmitting a certain amount of random bytes to the server.
5. Visual sensing: To characterize power consumed by the webcam, a sequence of frames is acquired.

Analysis of the energy consumption profile helps with the selection of hardware components for multimedia applications.

7.2.1 Microcontroller

Most WMSN computing subsystems are implemented as fully static CMOS MCU devices, which operate at frequencies ranging from very low (1 kHz) to a maximum speed of 1 MHz at 1.8 V, to 100 MHz at 5 V depending on the technology. The current drawn at 32 kHz is about 100 μA, when the MCU is running continuously, which is not sufficient to achieve multiyear battery life. In this case, it is necessary that the MCU be put into a power-savings mode, such as idle, sleep, or stop mode.

When the MCU is built entirely of clocked CMOS logic circuitry, the current consumption is a linear function of clock speed, with zero offset current. However, in modern mixed-mode MCUs, the flash-based type is packed with analog circuitry. Therefore, the total current consumed when the MCU is active is composed of two elemental components: static and dynamic current. The dynamic current consumption is the current change when clock frequency changes. The static current is the current that has different components that are independent of operating frequency; they are analog-block current, flash-module current, and leakage current.

Independent of the MCU clock frequency, most analog blocks have a significant current drain when they are powered. The flash memory typically draws current to power the array and read from the flash cell, which in some cases draws more current than the CPU itself, especially at low clock speed. The leakage current depends very much on the fabrication process technology.

In active mode, static current can reach 10 times as high as that in the power-savings mode. To minimize the active static current, the WMSN must operate at a low duty cycle, which means that faster speed results in more power savings. In addition, as lithography advances for low-voltage submicron technology, the dynamic current goes down, while the static leakage current tends to increases. Thus, for more advanced fabrication technologies, low duty-cycle operation is more beneficial. Table 7.1 shows the power consumption of the most popular CPU installed in standard sensor nodes [10].

7.2.2 Radio Communication System

Radio frequency (RF) communication requires modulation, band pass filtering, and multiplexing circuitry, which makes it more complex and power hungry. The path loss of the transmitted signal between two sensor nodes can be as high as the fourth-order exponent of the distance between them. Nevertheless, RF communication is preferred in most of the ongoing sensor network research projects, because the packets conveyed in sensor networks are small, data rates are low, and the frequency reuse is high due to short communication distances [11].

Several factors affect the power consumption characteristics of the communication system. These include the type of modulation scheme, data transfer rate, transmit power, and the operational duty cycle. Additionally, as device sizes shrink, wireless power consumption is becoming a dominant part of device power budget [12]. Communication energy consumption in both the transmitter and the receiver can be modeled as [13]

$$E_{tx} = (\alpha_t + \alpha_{amp}\, d^2)^* \, r \tag{7.1}$$

and

$$E_{rx} = (\alpha_r \, {}^* r) \tag{7.2}$$

Table 7.1 Power Consumption for Some Common CPUs

CPU	POWER SUPPLY (V)	POWER ACTIVE (mW)	POWER DOWN (μW)	SENSOR NODE
4-BIT CPU				
EM6603	1.2–3.6	0.0054	0.3	
EM6605	1.8–5.5	0.012	0.9	
8-BIT CPU				
ATtiny261V/461V/861V	1.8–5.5	*0.38 mA @ 1.8 V 1 MHz	*0.1	
PIC16F877	2.0–5.5	1.8		CIT
MC68HC05PV8A	3.3–5.0	4.4	485	
AT90LS8535	4.0–6.0	15	45	WeC Rene
ATmega163L	2.7–5.5	15		Rene2 Dot
ATMega103L	2.7–3.6	15.5	60	Mica IBadse
C8051F311	2.7–3.6	21	0.3	Parasitic
ATmega128L	2.7–5.5	26.7	83.15	Mica Mica2 Dot Mica2 BTnode
PIC18F452	2.0–5.5	40.2	24	EnOcean TCM
80C51RD+	2.7–5.5	48	150	RFRAIN
16-BIT CPU				
MSP430F149	1.8–6.0	3	15	Eyes.BSN
MSP430F1611	1.8–3.6	3	15	Telos
		1.5	6	SNoW[5]
MC68EZ326	3.3	60	60	SpotON
32-BIT CPU				
Atmel AT91 ARM Thumb	2.7–3.6	114	480	
Intel PXA271	2.6–3.8	193	1800	iMote2
Intel StrongArm SA-1100	3.0–3.6	230	25	WINS μAMPS

where E_{tx} is the energy to send r bits and E_{rx} is the energy consumed to receive r bits, α_t is the energy/bit consumed by the transmitter circuit, α_{amp} is the energy dissipated in the transmitter amplifier, α_r is the energy/bit consumed by the receiver circuit, and d is the distance that the message traverses, assuming a $1/d^2$ path loss. Table 7.2 shows power consumption of the most popular transceiver radio modules used by sensor nodes [14].

Table 7.2 Power Consumption for Some Common Radio Modules

TYPE	CLOCK (MHz)	Rx POWER (mA)	Tx POWER (mA/dBm)	POWER DOWN (μA)
LOW-POWER RADIO MODULES				
MPR300CB	916	1.8	12	1.0
SX1211	868–960		25/10	
TR1000	916	3.8	12.0/1.5	0.7
CC1000	315–915	9.6	16.5/10	1.0
MEDIUM-POWER RADIO MODULES				
nRF401	433–434	12	26/0	
CC2500	2400	12.8	21.6	
XE1205	433–915	14	33/5	0.2
CC1101	300–928	14.7	15	0.2
CC1010	315–915	16	34/0	0.2
CC2520	2400	18.5	17.4/0	<1.0
CC2420	2400	19.7	17.4/0	1.0
CC1020	402–915	19.9	19.9	0.2
CC2430	2400	19.9	19.9	
PH2401	2400	20	20	
nRF2401	2400	22	10/0	0.4
CC2400	2400	24	19/0	1.5
CC2530F32	2400	24	29/1	
RC1180	868	24	37/0	
LMX3162	2450	27	50	
STD302N-R	869	28	46.0	
MCI13191/92	2400	37	34/0	1
HIGH-POWER RADIO MODULES				
ZV4002	2400	65	65/0	140

Source: Stojcev, M.K. et al., *9th International Conference on Telecommunication in Modern Satellite, Cable, and Broadcasting Services*, © 2009 IEEE.

7.2.3 Sensing System

Sensor transducers convert physical quantities into electrical signals. According to the type of output they produce, sensors can be classified as analog or digital circuits. There exists a diversity of sensors that measure environmental parameters such as light intensity, temperature, humidity, proximity, magnetic fields, acoustic, etc. In general, passive sensors such as temperature sensors consume negligible power relative to other components of the sensor node. By comparison to passive sensors, active sensors such as imagers and proximity, pressure, flow control, and level sensors usually have an acquisition time that is longer than the transmission time, especially in the case of multimedia sensors such as CCDs and CMOS image sensors; accordingly, they consume more energy than communication [15,16].

Power consumption in WMSN sensors is mainly a result of signal sampling and conversion of physical signals, signal conditioning, and analog-to-digital conversion. Sensors such as acoustic and image sensors generally require high-rate and high-resolution power-hungry analog-to-digital converters. The power consumption of the converters can account for the most significant power consumption of the sensing system [17].

Direct power consumption is not the only factor affecting the selection of sensor for use in WMSNs; several other factors need to be considered. These factors include volume, suitability for power cycling, fabrication and assembly, compatibility with other components of the system, and packaging needs. Table 7.3 lists the power consumption of some common sensors.

Sensors can be divided into five groups, as shown in Table 7.3. The on/off sensors belong to the micropower group, with power consumption less than 1 mW. The second group, referred to as "low power," is characterized by power consumption less than 10 mW and a small amount of linear signal processing. Sensors from the medium-power group have a power consumption within the range of 10 to 50 mW and are realized with mixed circuits (analog and digital electronics). The sensors in the high-power group are enabled with some kind of dedicated signal processor, making the sensors of this group smart devices. The power consumption of this group is from 50 mW to 1 W. The last group, the ultra-high-power group, is characterized by consumption that is greater than 1 W. Due to higher power consumption in the last two groups, harvesting electronics are usually essential [18].

Table 7.4 provides an overview of the power consumption of multimedia sensors by comparing the power consumed by four classes of cameras, from very low-power, low-resolution cameras, to web cameras, to advanced, high-resolution cameras are available commercially. At the lowest end of the spectrum is the tiny Cyclops [19], which consumes a mere 46 mW and can capture low-resolution video. CMUcams [20] are cell-phone-class cameras with onboard processing for motion detection, histogram computation, and so on.

Table 7.3 Power Consumption for Some Common Sensors

SENSOR TYPE	SENSING	POWER CONSUMPTION (mW)
MICRO POWER		
SFH 5711	Light sensor	0.09
DSW98A	Smoke alarm	0.108
SFH 7741	Proximity	0.21
SFH 7740	Optical switch	0.21
ISL29011	Light sensor	0.27
STCN75	Temperature	0.4
LOW POWER		
TSL2550	Light sensor	1.155
ADXL202JE	Accelerometer	2.4
SHT 11	Humidity/ temperature	2.75
MS55ER	Barometric pressure	3
QST108KT6	Touch	7
SG-LINK(I000Ω)	Strain gauge	9
MEDIUM-POWER		
SG-LINK(350Q)	Strain gauge	24
iMEMS	Accelerometer	30
OV7649	CCD	44
2200/2600 series	Pressure	50
HIGH POWER		
TI50	Humidity	90
DDT-651	Motion detector	150
EM-005	Proximity	180
BES 516-371-S49	Proximity	180
EZ/EV-18M	Proximity	195
GPS-9546	GPS	198
LUC-M10	Level sensor	300
CP18.VL18.GM60	Proximity	350
TDA0161	Proximity	420
ULTRA HIGH POWER		
FCS-GL1/2A4-AP8X-H1141	Flow control	1250
FCBEX11D	CCD	1900/2800
XC56BB	CCD	2200

Source: Stojcev, M.K. et al., *9th International Conference on Telecommunication in Modern Satellite, Cable, and Broadcasting Services*, © 2009 IEEE.

Table 7.4 Power Consumption and Capabilities of Four Classes of Camera Sensors

MULTIMEDIA SENSOR	POWER USED FOR IMAGE CAPTURING	IMAGE CAPTURING CAPABILITY
Cyclops	42 mW	Fixed-angle lens, 352 × 288 at 10 fps
CMUcam	200 mW	Fixed-angle lens, 352 × 288 up to 60 fps
Webcam	200 mW	Auto-focus lens, 640 × 480 at 30 fps
High-end PTZ camera	1 W	PTZ lens, 1024 × 768 up to 30 fps

Source: Margi, C.B. et al., *Proceedings of 24th Brazilian Symposium on Computer Networks*, 2006.

Table 7.5 Average Power Requirements in Watts

SYSTEM STATE	POWER
Idle	1.473
CPU loop	2.287
Camera with CPU	3.049
Camera in sleep mode with CPU	1.617
Networking on with CPU	2.557
Camera, networking, CPU	4.280
Capture running	5.268
Sleep	0.058

Source: Soro, S., Heinzelman, W., *Advances in Multimedia*, pp. 1–22, 2009.

At the high end, webcams can capture high-resolution video at a full frame rate while consuming 200 mW, whereas pan–tilt–zoom cameras are retargetable sensors that produce high-quality video while consuming 1 W. An interesting example of power consumption for different tasks performed by a camera node is shown in Table 7.5 [21,22]. Each task has an associated cost and execution time.

7.2.4 Batteries System

From the system's point of view, a good microbattery should have the following features [23]:

- High energy density
- Rechargeable, in case the system has an energy harvester
- Small cell potential (0.5–1.0 V), such that digital circuits might take advantage of the quadratic reduction in power consumption with supply voltage

- Efficiently divided into series batteries to provide a variety of cell potentials for various components of the system without requiring voltage converters
- Large active volume-to-packaging ratio

Three small cell chemistries are currently dominating the application market for WMSNs: lithium–ion (Li-ion), nickel–metal hydride (NiMH), and lithium polymer (Li-polymer) [24]. Each of these battery types has its advantages and disadvantages for use in a node. The first step in selecting a cell for a node is to study the specific characteristics of each cell in terms of voltage, charging time, discharging rates, cycles, load current, and energy density. Following is a brief overview of the characteristics, advantages, and disadvantages of each of the three cell chemistries. The crucial battery parameters are given in Tables 7.6 and 7.7 [25].

Lithium–ion (Li-ion): The Li-ion battery has a nominal voltage of 3.6 V, 1000 duty cycles per lifetime, less than 1 C optimal load current,

Table 7.6 Battery Parameters

VOLTAGE	NOMINAL CELL VOLTAGE
Capacity	The amount of electrical charge that can be stored
Specific energy	The volume-related content, measured in energy/weight
Energy density	The volume-related content, measured in energy/volume
Internal resistance	Characterizes the ability to handle a specific load
Self-discharge	The internal leakage and aging effects
Recharge cycles	The number of charge cycles before performance degrades
Charging procedure	Type of charge circuit required

Source: Stojcev, M.K. et al., *9th International Conference on Telecommunication in Modern Satellite, Cable, and Broadcasting Services*, © 2009 IEEE.

Table 7.7 Battery Types

BATTERY TYPE	VOLTAGE (V)	ENERGY DENSITY (Wh/dm^3)	SPECIFIC ENERGY (Wh/kg)	SELF-DISCHARGE PER MONTH (%)
Lead–acid	2.0	60–75	30–40	3–20
Nickel–cadmium	1.2	50–150	40–60	10
Nickel–metal hydride	1.2	140–300	30–80	30
Lithium-ion	3.6	270	160	5
Lithium polymer	3.7	300	130–200	1–2

Source: Stojcev, M.K. et al., *9th International Conference on Telecommunication in Modern Satellite, Cable, and Broadcasting Services*, © 2009 IEEE.

an average energy density of 160 Wh/kg, a charging time of less than 4 hours, a typical discharge rate of approximately 10% per month when stored, and a rigid form factor. These characteristics make the Li-ion battery a good option when requirements include lower weight, higher energy density or aggregate voltage, and a greater number of duty cycles and when the price is not critical. To increase the voltage, Li-ion battery systems can be connected with up to seven series cells, resulting in a maximum aggregate voltage of 25.2 V [26].

Nickel–metal hydride (NiMH): The NiMH battery has a nominal voltage of 1.25 V, 500 duty cycles per lifetime, less than 0.5 C optimal load current, an average energy density of 100 Wh/kg, less than 4-hour charge time, a typical discharge rate of approximately 30% per month when in storage, and a rigid form factor. NiMH battery systems excel when lower voltage requirements or price sensitivities are primary considerations in cell selection. To increase the voltage, NiMH systems can be connected with up to ten series cells, resulting in a maximum aggregate voltage of 12.5 V [26].

Lithium polymer (Li-polymer): The Li-polymer battery has a nominal voltage of 3.6 V, 500 duty cycles per lifetime, less than 1 C optimal load current, an average energy density of 160 Wh/kg, less than a 4-hour charge time, a typical discharge rate of less than 10% per month when in storage, and a semi-rigid form factor. To increase the voltage, Li-ion cells can be connected with up to seven series cells, resulting in a maximum aggregate voltage of 25.2 V [26]. Li-polymer cells have a similar performance to Li-ion cells, but have the advantage of being packaged in a slightly flexible form as long as it remains flat when installed in a device.

7.3 Power Management Techniques

Various power management techniques are used to reduce the power consumption in different types of networks. Many of these techniques are used in common practice to reduce the power of devices in WMSNs. Doing so requires the use of a structured, interconnect-oriented design methodology at all layers, starting from the application layer all the way down to the physical layer of a networking protocol stack.

7.3.1 Application Layer

At the application layer, different techniques can be used to reduce the power consumption in a wireless device. In a load-partitioning technique, the application has the option to perform all of its power-intensive computations at the base station rather than locally. The wireless device sends a request for the computation to be performed and then waits for the result to be received. Another technique uses proxies to inform an application about battery power changes. Applications use this information to limit their functionality to provide only their most essential features and eliminate some unnecessary visual effects that accompany a process [27].

These techniques can be adapted to work with most applications that wish to support them. A number of other techniques also exist for specific classes of applications. Two such common applications are database operations and video processing. For database systems, techniques are able to reduce power consumed during data retrieval, indexing, and querying operations. In these cases, energy is conserved by reducing the number of transmissions needed to perform such operations.

For video-processing applications, energy is conserved by reducing the number of bits transmitted over the wireless channels using compression techniques. However, performing the compression itself may consume more power than that saved in transmission. Other techniques that slightly degrade video quality have been explored to reduce the power even further [28].

7.3.2 Transport Layer

All the various techniques used to conserve energy at the transport layer work by reducing the number of retransmissions necessary due to packet losses because of a faulty wireless link. In a wired network, packet losses indicate congestion, requiring back-off mechanisms to account for it. In wireless networks, however, packet losses can occur sporadically, which does not necessarily indicate the onset of congestion. With this knowledge in mind, TCP-Probing [29] and Wave and Wait protocols [30] have been developed as replacements for traditional TCP. They guarantee end-to-end data delivery with high throughput and low power consumption.

Congestion control is another important issue that should be considered in transport protocols. Congestion is an essential problem in WSNs. It not only wastes scarce energy due to a large number of retransmissions and packet drops, but also hampers the event detection reliability. Congestion in WMSNs has a direct impact on energy efficiency and application quality of service. Two types of congestion can occur in sensor networks [31]. The first type is node-level congestion that is caused by buffer overflow in the node, which can result in packet loss and increased queuing delay. Not only can packet loss degrade reliability and application quality of service, but it can also waste the limited node energy and degrade link utilization. In each sensor node, when the packet arrival rate exceeds the packet service rate, buffer overflow may occur. It is more likely to occur at sensor nodes close to the sink, as they usually carry more combined upstream traffic. The second type of congestion is link-level congestion related to the wireless channels, which are shared by several nodes using protocols such as CSMA/CD (Carrier Sense, Multiple Access with Collision Detection). In this case, collisions can occur when multiple active sensor nodes try to seize the channel at the same time. To avoid the negative aspects of congestion in WMSNs, congestion must be effectively controlled. Each congestion control solution consists of three important parts: congestion detection, congestion notification, and rate adjustment [32].

In traditional TCP protocol, congestion is detected at the end nodes based on a time-out or redundant acknowledgments. In general, link-by-link congestion detection in sensor networks has better performance than traditional end-to-end congestion detection using time-out or duplicate acknowledgment. Thus, in sensor networks, proactive methods are used, based on some form of congestion indicator. Different congestion indicators have been proposed, such as queue length [33], packet service time, or the ratio of packet service time to packet interarrival time at the intermediate nodes [34]. After detecting congestion, to prevent the negative aspects of congestion in the networks, the transport protocol needs to propagate congestion information from the congested node to the upstream sensor nodes or the source nodes that contribute to congestion. This can be done explicitly by sending a special control message to the other sensors, or implicitly by using a piggybacking technique in data packets.

When a node receives a congestion notification message, it will adjust its transmission rate using a rate control technique such as additive increase multiplicative decrease.

7.3.3 Network Layer

Power management techniques at the network layer level are responsible for performing power-efficient routing through a multihop network [35,36]. Despite the large volume of research activities and the significant progress made in recent years, routing in WMSNs still harbors many open issues that need to be resolved [37]. Because an idle receive circuit can consume almost as much power as an active transmitter, a good power-saving technique should permit as many nodes as possible to turn their radio receivers off most of the time. At the same time, it should forward packets between source and destination with almost the same delay as if all nodes were awake. This expectation implies that enough nodes should be awake to form a connected network.

The protocols used for the network layer are typically backbone-based, topology-control-based, or a hybrid of the two. In a backbone-based protocol, some nodes must remain active all the time to form a backbone, while others might sleep periodically. The backbone nodes will establish a path between all source and destination nodes in the network. This design means that paths that could operate without interference in the original network should be represented in the backbone. Therefore, any node in the network, including backbone nodes themselves, must be within one hop of at least one backbone node. The power savings is achieved by permitting nonbackbone nodes to sleep periodically, as well as by periodically choosing which nodes in fact make up the backbone. The algorithm for picking this backbone should be distributed, requiring each node to make a local decision. Furthermore, the backbone nodes should provide about as much total capacity as the original network, otherwise congestion may increase.

Each node in the network periodically makes local decisions on whether to sleep or stay awake as a coordinator to participate in the forwarding backbone structure. A node volunteers to become a coordinator if it discovers, using information gathered from local broadcast messages, that two of its neighbors are not communicating with each other directly or through existing coordinators. This provision will preserve the capacity of the network. To keep the number of redundant coordinators low, each node delays announcing its willingness to become a coordinator by a random time interval. The length of this time interval considers two factors: the amount of remaining battery charge and the number of pairs of neighbors that can connect together. This role will rotate among all nodes, ensuring, with high probability, a capacity-preserving connected backbone at any point in time, where nodes lean to consume power at about the same rate.

Backbone-based protocols, such as ASCENT [38] and SPAN [39], utilize local information to assess the connectivity of a node with its neighbors and then decide whether or not the node should stay active to join a communication backbone. These protocols concentrate on maintaining continuous connectivity of the network and are best suited for high data rate ad hoc multihop networks.

Topology-based routing protocols can achieve power savings differently. Their goal is to reduce the transmission power by making all nodes operate with their lowest possible transmission power, such that the network just remains connected. In a homogeneous network, this design would mean that the all nodes adjust their transmission powers so that they are just within the range of their nearest one-hop neighbor. In heterogeneous networks, the transmission powers will be adjusted based on the needs of that network. A summary of the different types of existing topology-based protocols is shown in Figure 7.1.

Certain location-based topology control protocols try to use the topology of the network to provide the most power-efficient communication path possible. These protocols produce a localized power-aware routing mechanism for the network. In some cases, providing

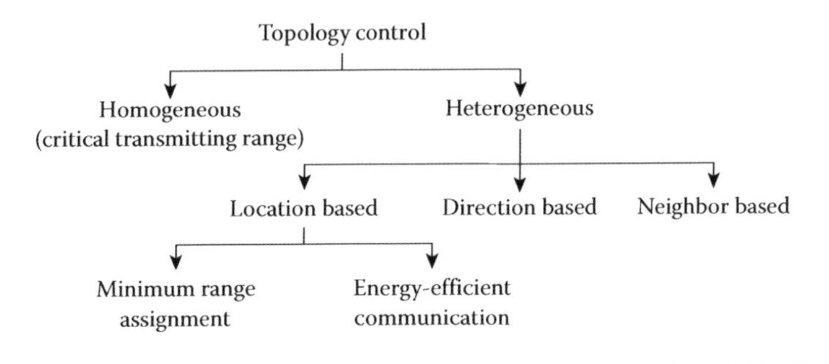

Figure 7.1 Topology-based routing protocols.

such a path means taking a larger number of hops through the network than that taken by direct transmission from one node to another. It is more energy efficient to transmit over several short distances than it is to transmit over a few long distances [40]. Short distances require less energy to transmit across and enable better signals, resulting in fewer retransmissions due to packet loss. Other power-aware routing protocols using connected dominating sets elect fewer coordinators because they actively prevent redundant coordinators using randomized slotting and damping [41].

Introducing new sensor nodes and allowing the network to self-organize and learn often offer better solutions. Such a plan allows the network to perform better in a dynamic environment according to its acquired knowledge [42]. For such dynamic networks, nodes cannot know *a priori* the optimal route to other nodes, because the paths keep changing as nodes move, enter, or leave the network. Therefore, the network protocol coordinates the discovery and tracking of routes in the network. This discovery and tracking, however, consumes energy because it requires communication between nodes. With the low data rates and fast dynamics of some nodes, the network discovery and maintenance overhead may dominate the power consumed for data transmission itself [43]. Because environmental conditions and user constraints can vary over time, the use of static algorithms and protocols can result in less than optimal energy consumption. Thus, wireless sensors must allow adaptation of underlying hardware by higher level algorithms. By giving upper layers the opportunity to adapt the hardware in response to changes in the state of the system, the user's quality constraints and the energy consumption of the node can be better controlled [44].

Finally, transmission power control schemes might be combined with backbone-based ones to produce a hybrid of both, as the benefits of these protocols can be achieved simultaneously.

7.3.4 Data Link Layer

Automatic Repeat Request (ARQ) and Forward Error Correction (FEC) schemes are the most common techniques used to conserve energy at the data link layer to reduce the transmission overhead. These schemes aim to reduce the number of packet errors at a receiving node.

By enabling ARQ, a router can directly request a retransmission of a packet from its source without first having to require the receiver node to detect if a packet error has occurred. Power-saving results have shown that occasionally it is more efficient to transmit at a lower transmission power while sending multiple ARQs than to transmit at a high transmission power and achieve better throughput. Adding FEC to ARQ codes will result in a reduction of the number of retransmissions necessary at a lower transmission power, which increases energy efficiency [27].

Another power management technique applied at the data link layer based on a packet-scheduling protocol [45] allows multiple packet transmissions to occur back to back. It might be possible to reduce the power associated with sending each packet individually. To announce the presence of sent packets on the radio channel, preamble bytes need to be sent for the first packet only, while all subsequent packets will follow this announcement. A packet-scheduling algorithm might also reduce the number of retransmissions necessary only if the packet is sent during a scheduled time when its destination is able to receive packets. By reducing the number of retransmissions necessary, the overall power consumption will consequently be reduced.

7.3.5 MAC Layer

Idle listening—the time spent listening while waiting to receive packets—is a significant cost. The energy cost for a node in idle mode is approximately the same as in receive mode. Even when no communication is taking place, a considerable amount of energy is spent searching for the next packet. In many application scenarios, the energy spent while waiting for a transmission can represent more than 90% of a node's total energy budget. It has become clear that to reduce power consumption in radios, the radio must be *turned off* during idle times (sleep) [46–47].

Power-saving techniques at the MAC layer are primarily concerned with sleep-scheduling protocols. Such protocols switch a radio's power on and off to reduce the effects of this idle listening. They are used to wake up the radio whenever it is required to transmit or receive packets; otherwise, it will sleep.

Battery-Aware MAC (BAMAC) protocol [48] decides which node should send next based on the battery level of all surrounding nodes in the network. Battery level information is attached with each transmitted packet, and decisions about sending packets from individual nodes will be based on this information.

To complete the above procedure, several challenges must be overcome. First, efficient mechanisms are needed to selectively activate sleeping nodes with the most remaining energy according to the appropriate positions around the target. Second, the activated nodes need to cooperate to distribute the required signal processing tasks with a certain quality of service, aggregate the results, and route the final decision to the base station. Third, network management can be complicated when different nodes have varying energy levels, processing capabilities, and sensing modalities [49].

Sleep-scheduling protocols are divided into two categories: synchronous and asynchronous [50,51]. Synchronous sleep-scheduling policies rely on a synchronized clock between all nodes in a network. Senders and receivers each know when the other should be on and only send packets to one another during these slotted time intervals; otherwise, they go to sleep. Slotted time division has a natural structure that leaves traffic uncorrelated and provides end-to-end fairness with high power efficiency. Global schedules can be generated with reserved bandwidth from source to sink and make it clear when to turn the radio on and off locally. Slotted time-division schemes have static global schedules that require centralized control and very precise time synchronization [52].

PAMAS (Power Aware Multi-Access Signaling) [53,54] is a synchronous solution that has proven to be very effective in reducing the power consumed by nodes by detecting when a packet on the channel is sent for someone else and putting themselves to sleep. This approach is suitable for radios in which processing a received packet is expensive compared to listening to an idle radio channel. PAMAS can be combined with some of the other sleep-scheduling protocols to produce even more power savings.

In the AFECA protocol [55], each node listens to the transmissions on the channel and maintains a count of the number of nodes within radio range. A node then switches between listening and sleeping, with randomized sleep times proportional to the number of nearby nodes with a net effect of maintaining the number of listening nodes roughly constant, regardless of node density. As the node density increases, more power can be saved. To be conservative, AFECA tends to make nodes listen even when they could be asleep if a node does not know whether it is required to listen to maintain connectivity. Setting the on/off periods based on application hints reduces both power and delay [56].

Asynchronous sleep scheduling, on the other hand, does not rely on any node clock synchronization at all. According to the MAC protocol in use, nodes can send and receive packets whenever they like. Nodes wake up and go to sleep periodically in the same way they do for synchronous sleep scheduling.

However, it has been shown that in some situations poor interaction between high-level power-saving techniques and low-level communication protocols might lead to an increase in power consumption when using these mechanisms [56]. Therefore, there must be a way to ensure that receiving nodes are awake to hear other nodes' transmissions. Precise control over the power state of the radio allows the protocol to turn off the radio between each sample. To synchronize the starting point of an incoming data stream between the transmitter and receiver, preamble bytes are sent by a packet. Once detected, the preamble will cause the receiver to search for the pending start symbol. The duty cycle of the receiver becomes proportional to the length of this preamble.

A sufficient number of extra preamble bytes are sent per packet to assure that a receiver has the chance to synchronize with it at some point. In the worst case, a packet will be transmitted just as its receiver goes to sleep, and preamble bytes will have to be sent for a time equal to the receiver's sleep interval. However, the frequency with which a receiver checks for a wake-up signal controls the amount of time that it takes for the network to wake up. Once the receiver wakes up, it will synchronize to these preamble bytes and stay on until it completely receives the packet. This protocol optimization trades power consumption by the sender for power consumption by the receiver, because the sender must transmit longer but the receiver can sample the radio channel less frequently. The optimal ratio is dependent on the communication patterns of the application.

The receiver overhead can be reduced arbitrarily at the expense of bandwidth, latency, and transmission overhead depending on application-specific goals. Moreover, based on network activity with flexible communication protocols, an application can change the protocol at runtime by exploiting the ability to tailor protocols to application-specific criteria [52].

The Low Power Listening (LPL) [57] asynchronous sleep-scheduling protocol is quickly becoming the effective standard for sleep-scheduling policies. LPL operates in a similar fashion to any other asynchronous sleep-scheduling protocol, but with one key difference. LPL turns the radio on to check for an incoming packet through the channel very quickly and reliably so that it can go back to sleep immediately afterward. The time between each of these checks is known as a check interval. LPL only achieves significant power savings if many check intervals are allowed to pass before a packet is actually detected on the channel, which makes LPL ideal for the low data rate environment in WSNs.

Another asynchronous technique known as remote access switch is used to wake up a receiver only when data are being sent for it. A low-power radio circuit is always running to detect a certain type of activity on the channel. The circuit wakes up the rest of the system for reception of a packet only when this activity is detected. However, a transmitter has to know what type of activity needs to be sent on the channel to wake each of its receivers up [35].

An event starts when a sensor node picks up a signal above a predetermined threshold power. Every node is assumed to have a uniform radius of visibility r, while in real applications the terrain might influence the visible radius. An event, such as a localized change in parameters in an environment-monitoring application, can be static or can propagate, such as signals generated by a moving object.

In general, events have a characterized distribution in space and time. There are three distinct classes of events:

- The event occurs as a stationary point.
- The event propagates at a fixed velocity.
- The event propagates at a fixed speed but in a random direction.

The processor must watch for preprogrammed wake-up signals. Prior to entering the sleep state, the CPU programs these signal conditions. The node must be able to predict the arrival of the next event to wake up on its own.

A pessimistic strategy results in some events being missed, while an optimistic prediction might result in the node waking up unnecessarily. There are two possible approaches:

1. Disallow the state completely, which will result in events being missed because the node was not alerted. If the sensing task is critical and events cannot be missed, this state must be disabled.
2. Disallow the state selectively, resulting in events being missed because the node was not alerted. This technique can be used if the events are spatially distributed and not all critical.

Both random and deterministic approaches can be used. In the clustering protocol, the cluster heads can have a disallowed state, whereas normal nodes can transit to this state, which makes the scheme more homogeneous. Every node that satisfies the sleep-threshold condition for the selectively disallowed node can enter sleep with a system defined by a probability for a certain time duration.

The advantage of this algorithm is that an energy-efficiency compromise can be made with event detection probability. By increasing this probability, the probability of missed events will increase, while the system energy consumption will be reduced, and vice versa. Hence, the overall shutdown policy is controlled by two implementation-specific probability parameters.

The protocols described here explain only some of the main sleep-scheduling protocols that have been developed to date. They do provide a good indication of the different domains to which variable sleep-scheduling protocols are most applicable.

Unlike for the energy-efficient routing protocols, it does not make sense to have a hybrid sleep-scheduling protocol based on these two techniques. The energy savings achieved using each of these techniques varies from system to system and application to application. Efforts are being made to define exactly when each technique should be used, as one technique is not better than the other in this sense.

7.3.6 Physical Layer

Techniques can be implemented at the physical layer not only to preserve energy but also to generate it. Proper hardware design techniques can decrease the level of leak currents in an electronic device due to parasitic capacitance to almost nothing [58], resulting in a longer lifetime for these devices, because they consume less power while idle. Variable clock CPUs, CPU voltage scaling, flash memory, and disk spin-down techniques can also be used to further reduce the power consumed at the physical layer [35].

Energy-harvesting techniques allow a device to actually gather energy from its surrounding environment. Ambient energy is all around in the form of vibration, strain, inertial forces, heat, light, wind, magnetic forces, and so on [59]. Energy-harvesting techniques allow this energy to be harnessed and either converted directly into usable electric current or stored for later use within an electrical system.

In the next sections, the latest technological advances in both low-power design and energy-harvesting techniques will be introduced.

7.4 Dynamic Power Management of Very Large Scale Integrated Circuit (VLSI) Systems

Power awareness can be enhanced by applying systematic techniques to VLSI systems at several levels of the system hierarchy: multipliers, register files, digital filters, dynamic voltage scaled processing, and data-gathering wireless networks. The power awareness of these systems can be significantly enhanced, leading to increases in battery lifetimes.

A related motivation for power awareness is that a well-designed system must gracefully decline in quality and performance as the available energy resources are depleted. In this context, making a scalable system refers to enabling the user to compromise the performance of the system parameters by hard-wiring them. Scalability allows the end user to implement operational policy, which often varies significantly over the lifetime of the system.

While the above argues for power awareness from a user-centric and user-visible perspective, one can also motivate this paradigm in more fundamental, system-oriented terms. With burgeoning system complexity and the accompanying increase in integration, there is more diversity in operating scenarios than ever before. Even if there is little explicit user intervention, there is an imperative to track operational diversity and scale power consumption accordingly. This necessity naturally leads to the concept of power awareness. For instance, the embedded processor that decodes the video stream in a portable multimedia terminal can display tremendous workload diversity depending on the temporal correlation of the incoming video bit stream. Hence, even if the user does not change the quality criteria, the processor must exploit this operational diversity by scaling its power as the workload changes.

Architecture and VLSI technology trends point in the direction of increasing energy budgets for register files. The key to enhancing the power awareness of register files is the observation that over a typical window of operation, a microprocessor accesses a small group of registers repeatedly, rather than the entire register file. More than 75% of the time, no more than 16 registers are accessed by the processor in a 60-instruction window. Equally importantly, there is strong locality from window to window. More than 85% of the time, fewer than five registers change from window to window.

The number of registers the processor typically needs over a certain instruction window is considered a scenario. The smaller files have lower costs of access because the switched bit-line capacitance is lower. Hence, from a power awareness perspective, over any instruction window, as small as possible a file is used.

There are significant motivations for investigating power-aware filters. As an example, consider the adaptive equalization filters that are ubiquitous in communications application-specific integrated circuits. The filtering quality requirements depend strongly on the channel conditions (line lengths, noise, and interference), the state of the system (training, continuous adaptation, freeze, etc.), the standard dictated specifications, and the quality of service desired.

All these considerations lead to tremendous scenario diversity, which a power-aware filtering system can exploit. This is because using all available time allows the frequency of the processor to be lowered, which in turn allows scaling down of the voltage, leading to significant energy savings. In terms of the power awareness framework, a scenario would be characterized by the workload. The point systems would be processors designed to manage a specific workload. As the workload changes, we would ideally want the processor designed for the instantaneous workload to execute it. It is clear that implementing such an ensemble spatially is meaningless and must be done temporally using a dynamic voltage scaling (DVS) system.

Increased levels of integration and advanced low-power techniques are enabling dedicated wireless networks of sensor nodes. Replacing high-quality sensors with such networks has several advantages, including robustness, fault tolerance, and autonomous operation for years [60].

Traditionally, energy-efficient VLSI design has been focused on low-power techniques. As the issue of energy efficiency becomes more pervasive, the policy of using the bare minimum of energy will face different challenges: semiconductor technology, circuit design, design automation tools, system architecture, operating system, and application design [61].

7.4.1 On-Chip Power Management

Interconnect wires account for a significant fraction of the power consumption in an integrated circuit that can reach up to 50% [62], which is expected to grow in future, making on-chip interconnects crucial. As the technology scales to the nanometer regime, the delay and energy consumption of global interconnect structures will be a major bottleneck for the system-on-chip (SOC) design [63,64].

A suggested approach is to use an on-chip router-based interconnect architecture, as shown in Figure 7.2 [65], similar to that adopted at the board level for interconnecting components in a wireless multimedia node [66]. In theory, the router could be a fully connected crossbar. However, every component on the chip will not need to talk

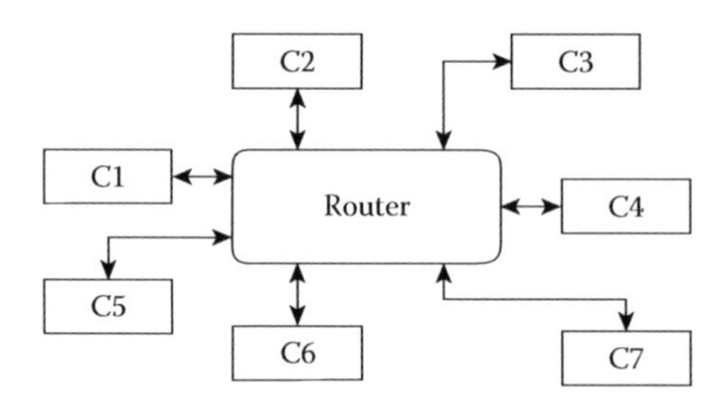

Figure 7.2 A router-based communication architecture.

to all other components; hence, it might be optimized for reduced complexity. Furthermore, optimizations can be used to reduce latency and improve crossbar utilization [67]. Other advantages of the router-based architecture are [65] as follows:

Components being isolated from each other leads to reduced capacitive loading during data transfers, which results in lower power consumption.

The increased parallelism enabled by the router results in a higher throughput, which can be further improved through the use of frequency/supply voltage scaling or other power-performance-improving techniques.

As the number of components present in the system increases, the power benefits of router-based communication architecture also increase [66]. Adopting router-based communication architecture will help future SOCs, which are expected to have a large number of components [68], which will potentially lead to significant power savings.

Researchers have just begun to investigate the merits and demerits of such an approach by analyzing and modeling the power consumption of routers and switches [69,70]. Efforts are also underway to develop modeling and simulation frameworks for on-chip communication architectures [71]. Such frameworks will enable SOC designers to quickly and efficiently explore the communication architecture design space.

7.4.2 Sensing Power Management

To reduce power consumption for the WMSNs containing power-consuming sensors, two approaches are considered: duty cycling and adaptive sensing. Duty cycling consists of powering the sensing system on only for the time needed to obtain a new set of samples, while switching it off immediately afterwards, provided that the dynamics of the sensed phenomenon are time invariant and known in advance. For fixed sampling, the rate should be computed *a priori*, which might have oversampling, inducing, in turn, waste of power. A better approach would require adopting a dynamically adaptive sensing strategy that tracks the real dynamics of the process. By reducing the number of samples, an efficient sensing strategy will also reduce the amount of data to be processed and transmitted to other network nodes. To provide effective handling of the duty-cycle issue, some aspects of the sensor drivers for the operating system must be considered. Failing to consider these aspects might result in invalid acquired data and power consumption larger than that associated with the traditional continuous powering mode. Those aspects that impact the power-managed sensor are defined by a set of functional characteristics, mainly wake-up latency and break-even cycle. The wake-up latency is defined as the time required by the sensor to produce a correct value once activated. It is clear that the active time of the sensor (t_{on}) needs to be long enough for the sensor to wake up (t_{wakeup}) and to obtain the measured information ($t_{acquire}$) [72]:

$$t_{on} \geq t_{wakeup} + t_{acquire} \tag{7.3}$$

The break-even cycle is defined as the rate at which the power consumption of a power-managed node equals that of non-power-managed node. This value is in inverse proportion to the overhead power consumption introduced by the nonideal on/off sensor transition, which represents the highest sampling rate for which applying power management is worth.

Adaptive sensing can be implemented by developing three main different approaches: hierarchical sensing, adaptive sampling sensing, and model-based active sensing, as indicated in Figure 7.3.

1. Hierarchical sensing techniques assume that multiple sensors are installed on the sensor nodes to measure the same physical quantity, each characterized by its own accuracy and

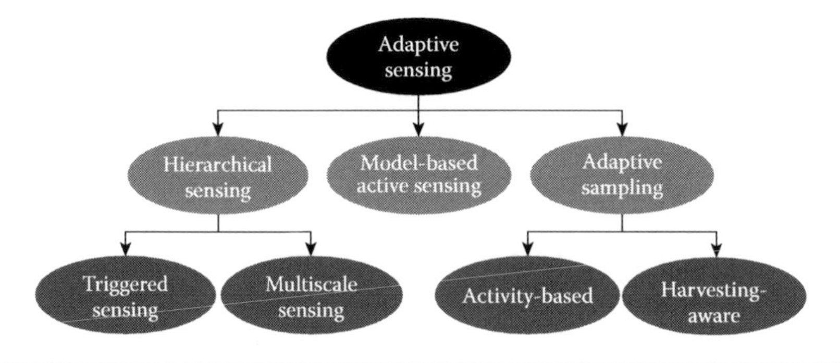

Figure 7.3 Classification of adaptive sensing strategies. (From Jeličić, V., Qualifying doctoral examination, University of Zagreb, Croatia, 2011.)

power consumption. Simple sensors are usually energy-efficient, but provide very limited resolution, whereas complex sensors can give a more accurate characterization of the sensed phenomenon at the expense of higher power consumption. At first, low-power sensors are activated to provide a coarse-grained characterization of the sensing field or to trigger an event, then accurate but power-hungry sensors can be turned on to provide measurements to improve the original coarse description. Two main approaches are used for hierarchical sensing: triggered sensing and multiscale sensing.

   a. Triggered sensing [73]: Activation of the more accurate and power-consuming sensors after the low-resolution ones detect some activity within the sensed area is referred to as "triggered sensing." A central node, which supervises all the activities of the WMSN, is endowed with a triggering system.
   b. Multiscale sensing [74]: A different use of hierarchical sensing consists of identifying areas within the monitoring field that need a higher resolution observation. Such areas can be identified by relying on a coarse-grained description of the field from lower accuracy sensors and activating additional high-resolution ones only in areas where it is requested to have accurate acquisitions.

2. Adaptive sampling techniques are aimed at dynamically adapting the sensor sampling rate by exploiting spatial and temporal correlations among acquired data and the available energy whenever the sensor node is able to harvest energy from the environment as follows:
   a. Activity-driven adaptive sampling: Activity-driven adaptive sampling exploits both temporal and spatial correlation among the acquired data.
   b. Harvesting-aware adaptive sampling: Harvesting-aware adaptive sampling techniques develop knowledge about the residual and predicted energy coming from the harvester module at the unit level to optimize its power consumption. The approach requires developing models able to characterize the progress of energy availability over time and the energy consumption of sensor units.

3. Model-based active sampling [75] consists of building a model of the sensed phenomenon on top of an initial set of sampled data. Once the model is available, instead of sampling the quantity of interest, the next data can be predicted by the model, hence saving the data-sensing power consumption. The model needs to be updated or reestimated whenever the requested accuracy is no longer satisfied to adopt the new dynamics of the physical phenomenon under observation. Correlation-based sampler selection is performed at each cluster head in order to determine the nodes that capture the best spatial and temporal correlations among the other sensor readings [76].

7.4.3 Low-Power System Design

Currently, most WMSN components are fabricated using CMOS technology. Intelligent, wireless microsensor node technology, based on commercial, low-cost CMOS fabrication and bulk micromachining, has demonstrated the capability to have multiple sensors, electronic interfaces, control, and communication in a single device. However, they face challenges in the form of the requirements for power consumption and the complete integration of a CMOS RF transceiver [77].

The main reason for the bias toward CMOS is that this technology is cost-efficient and inherently consumes less power than other technologies. The dominant factor of power consumption in CMOS is *dynamic*. A first-order approximation of the dynamic power consumption P_d of CMOS circuitry is given by the following formula [61]:

$$P_d = C_{eff} V^2 f \tag{7.4}$$

where C_{eff} is the effective switch capacitance, V is the supply voltage, and f is the frequency of operations. The power dissipation arises from the charging and discharging of the circuit node capacitance found on the output of internal circuit capacitances. C_{eff} combines two factors: C, the capacitance being charged and discharged, and the *activity weighting* α, which is the probability that a transition will occur.

$$C_{eff} = \alpha C \tag{7.5}$$

Therefore, lower level power consumption can be decreased by reducing the supply voltage, the capacitive load, or the switching frequency. Good system design attempts to make a system optimal for a certain application and environment, which takes into consideration various parameters such as supply voltage and clock frequency. However, energy efficiency in WMSNs is not a one-time design problem, to be solved during the design phase. Rather, it requires frequent adaptations to the system so that it can fulfill requirements in terms of a general quality of service model. This multidimensional design space offers a wide range of possible optimization.

7.4.4 Dynamic Voltage Scaling

It is well known that processor workloads can vary significantly, and it is highly desirable for the processor to scale its energy with the workload. The dynamic computing slowdown factor procedure is to alter the supply voltage and operating frequency of the system dynamically during task scheduling in accordance with recent tasks and execution history.

Although shutdown techniques can yield substantial energy savings in idle system states, additional energy savings are possible by optimizing the sensor node performance in the active state. DVS is an effective technique for reducing CPU energy. Simply reducing the operating frequency during periods of reduced activity results in linear decreases in power consumption, but does not affect the total energy consumed per task. Reducing the operating voltage implies greater critical path delays, which in turn compromise peak performance.

DVS is the active adjustment of the supply voltage in conjunction with the clock frequency in response to fluctuations in a processor's utilization. Peak performance is not always required and, therefore, the processor's operating voltage and frequency can be dynamically adapted according to the instantaneous processing requirement, which results in a nearly quadratic savings in energy and reduces leakage current. A voltage scheduler, running in tandem with an operating system's task scheduler, can adjust voltage and frequency in response to *a priori* knowledge or predictions of the system's workload. DVS has been successfully applied to custom chip sets.

DVS can be implemented using a DC–DC converter circuit with a dynamically digitally adjustable voltage that delivers power to a microprocessor core and is controlled by a multithreaded, power-aware operating system, as shown in Figures 7.4 and 7.5. A buck regulator composed of discrete components is driven by a commercial step-down switching regulator controller [78]. This controller is programmed with a 5-bit digital value to regulate 1 of 32 voltages between 0.9 and 2.0 V. The operating system running on the microprocessor commands the core voltage as a 5-bit digital value that is passed to the regulator controller. External programmable logic between microprocessor and the regulator controller prevents the regulator from delivering a voltage beyond the microprocessor core's rated maximum.

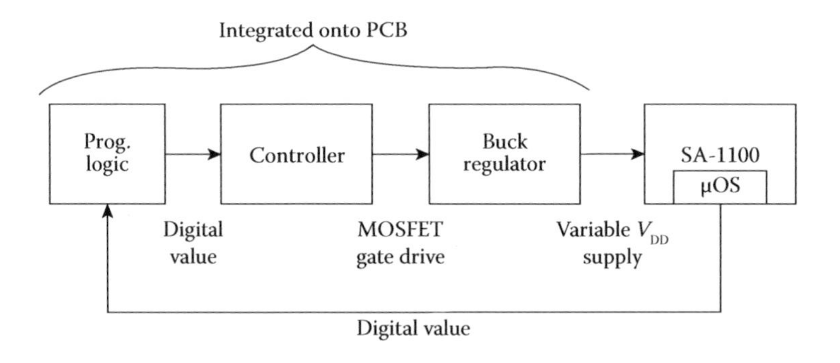

Figure 7.4 Overview of the adjustable DC–DC converter. (From Hac, A.: *Wireless sensor network designs*, 2003. Copyright Wiley-VCH Verlag GmbH & Co. KGaA.)

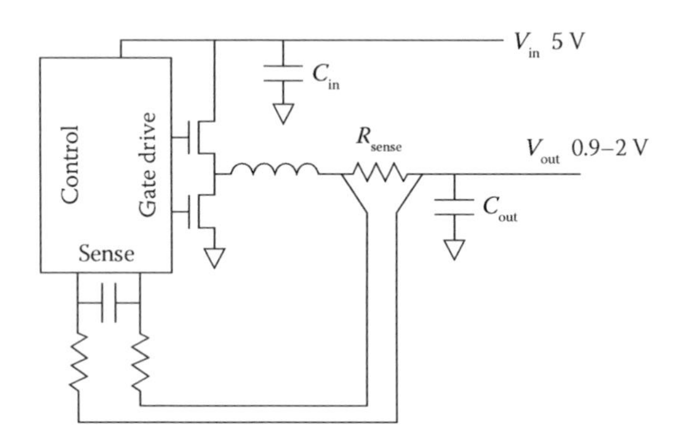

Figure 7.5 Simplified schematic for the buck regulator. (From Hac, A.: *Wireless sensor network designs*, 2003. Copyright Wiley-VCH Verlag GmbH & Co. KGaA.)

In the implementation, the microprocessor operation system monitors load on the processor and adjusts the clock frequency and supply voltage together to meet throughput requirements imposed by the tasks. Though the majority of throughput requirements and real-time deadlines on a sensor network are known *a priori*, more sophisticated load prediction algorithms may be needed for more optimal voltage scheduling for nondeterministic workload.

The rate at which DVS is carried out also has a significant bearing on performance and energy. A low update rate implies greater workload averaging, which results in lower energy use. The update energy and performance cost is also amortized over a longer time frame. Conversely, a low update rate also implies a greater performance hit, because the system will not respond to a sudden increase in workload.

7.4.5 Lifetime Prolongation Evaluation

To evaluate the scheduling scheme in terms of power conservation, we compare the cooperative scheduling scheme with a single-tier network or one tier of a multitier architecture consisting of N nodes monitoring without coordination among them, in which, nodes are awakened at a time period of T [79,80]. We note that the evaluation is over the sensing subsystem and that the radio subsystem (i.e., transmission and reception of packets) is not taken into account.

The energy (E) consumed in the network for object detection by N nodes during a duty-cycle interval of T in the noncollaborative scheduling is

$$E = N \cdot (T_{\text{sleep}} \cdot P_{\text{sleep}} + E_{\text{w\_up}} + E_{\text{cap}} + E_{\text{detect}}) \tag{7.6}$$

where T_{sleep} and P_{sleep} are the period and power consumption for a node in sleep mode. $E_{\text{w\_up}}$, E_{cap}, and E_{detect} are the energies consumed in waking up a node, capturing a picture, and performing object detection, respectively.

Consider the cooperative scheduling algorithm in a clustered tier/network; both the interval between waking up consecutive nodes in the same cluster and the period during which a given node is awakened, are functions of the size of the cluster that the nodes belong to.

Note that, in large clusters, T_{interval} is small and thus cluster duty-cycle frequency is increased. Moreover, a larger number of nodes in the cluster takes a longer period T_P to awaken a given node of the cluster and thus enhances the power conservation in the cluster members. Assuming an average cluster size for all clusters in the tier/network, T_P will be

$$T_P = \frac{T \cdot \mu C_{\text{size}}}{\mu C_{\text{size}} - \gamma \cdot (\mu C_{\text{size}} - 1)} \tag{7.7}$$

where T is the base period for waking nodes in the base uncoordinated tier, C_{size} is the size of the cluster, μC_{size} is the average cluster size, and γ is the clustering scale.

Figure 7.6 shows the evolution of T_p normalized by T (i.e., $\mu C_{\text{size}}/\beta$) for several node densities and γ [18]. The factor β represents the increment of area that the cluster senses with respect to an individual sensor.

Consequently, the total average energy consumption by nodes for object detection in the coordinated tier during T_P will be

$$E_P = E + N \cdot P_{\text{sleep}} \cdot (T_P - T) \tag{7.8}$$

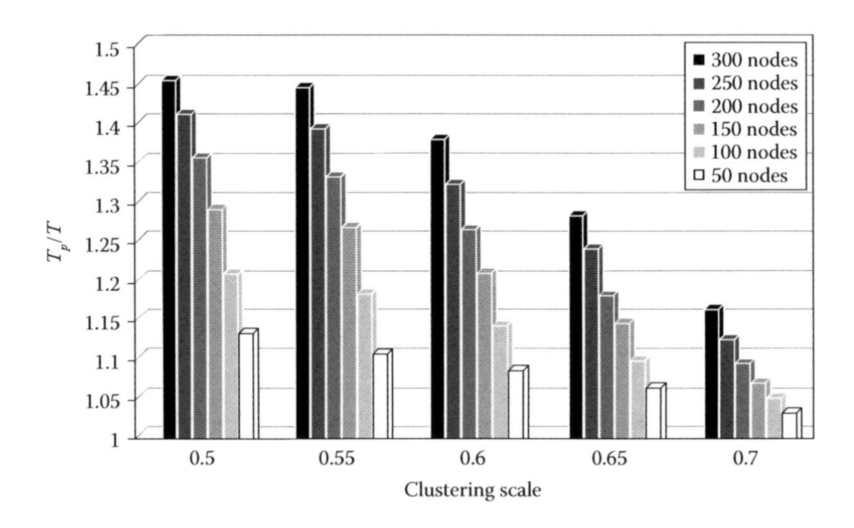

Figure 7.6 T_P/T for several node densities and clustering scales. (From Alaei, M., and Barceló, J., *Wireless communications and networks: Recent advances*, Rijeka, Croatia, InTech, 2012.)

7.5 Energy Harvesting

Energy-harvesting techniques, also referred to as "energy scavenging," extract energy from the environment where the sensor itself exists, offering another important means to prolong the lifetime of sensor devices. Traditionally, energy has been harvested through solar light, background radio signals, thermoelectric conversion, vibrational excitation, and the human body [81]. Solar panels are made up of *photovoltaic* cells that convert sunlight directly into electrical current, coupled with super capacitors and rechargeable batteries [82]. However, the primary disadvantage of solar panels is that they are large and require continuous sunlight to work. In most wireless networking situations, it is not practical to be limited by such constraints. Solar panels achieve only about 16–18% efficiency [66].

Super capacitors can also be used in these systems to effectively lower the impedance of a battery of the energy-harvesting system to allow larger peak currents or to store charge from the energy harvester to compensate for lulls, such as nighttime for a solar cell. Super capacitors can store up to 10 mJ/mm^3, which is less than 1% of the energy density of lithium cells [24].

Most recently, technology using piezoelectric materials has been used for power generation. Piezoelectricity is an electric current created by subjecting certain types of crystal to mechanical stress. The primary advantage of piezoelectric materials over solar panels is that they are small, do not require access to direct sunlight, and operate with about a 70% mechanical-to-electrical transduction efficiency.

For collecting energy from background radio signals, an electric field of 1 V/m yields only 0.26 1 W/cm^2, in contrast to 100 1 W/cm^2 produced by a crystalline silicon solar cell exposed to bright sunlight [3]. Electric fields of intensity of a few V/m are only encountered when it is close to strong transmitters. Another practice of broadcasting RF energy deliberately to power electronic devices is severely constrained by legal limits set for health and safety concerns.

Unlike thermoelectric conversion, which may not be suitable for wireless devices, harvesting energy from vibrations in the surrounding environment can provide another useful power source. Vibration-based magnetic power generators derived by moving magnets or coils may yield power that ranges from tens of microwatts when

Table 7.8 Power Output from Various Energy-Scavenging Technologies

HARVESTING TECHNOLOGY	POWER DENSITY
Solar cells—direct sun	15 mW/cm^2
Solar cells—cloudy day	0.15 mW/cm^2
Solar cells—indoors	0.006 mW/cm^2
Solar cells—desk lamp <60 W	0.57 mW/cm^2
Piezoelectric—shoe inserts	330 μW/cm^2
Vibration—microwave oven	0.01–0.1 mW/cm^2
Thermoelectric—10°C gradient	40 μW/cm^2
Acoustic noise—100 dB	9.6–4 mW/cm^2
Passive human-powered system	1.8 mW^2
Nuclear reaction	80 mW/cm^3 1E6 mWh/cm^3

Source: Stojcev, M.K. et al., *9th International Conference on Telecommunication in Modern Satellite, Cable, and Broadcasting Services*, © 2009 IEEE.

generated by microelectromechanical system technologies to over a milliwatt for larger devices. Other vibration-based microgenerators are relying on charged capacitors with moving plates and, depending on their excitation and power conditioning, yield power in the order of 10 1 W [3]. However, recent microgenerators [83] have yielded up to 800 μW/cm^3 from machine-induced stimuli, which is orders of magnitude greater than the power provided by currently available microgenerators [81]. Hence, this seems to be a promising technology for small battery-powered devices. Table 7.8 shows power outputs for typical energy-scavenging devices.

Although these techniques may provide an additional source of energy and help prolong the lifetime of sensor devices, they yield power that is several orders of magnitude lower compared to the power consumption of state-of-the-art multimedia devices. Hence, they may currently be suitable only for very low duty-cycle devices.

7.5.1 Energetic Sustainability

A workload can be called energetically sustainable if, at each node, the power required for processing data packets and routing them to base stations is sustained completely by the power harvested from the environment.

The maximum energetically sustainable workload (MESW) is defined as a workload that can be energetically sustained by each node involved in packet processing and routing and that cannot be incremented without violating the energetic sustainability at some nodes. In the case of continuous uniform monitoring applications, the MESW is defined as the maximum rate at which data packets are produced by all sensors and delivered to base stations. Both the MESW of the routing algorithm under study and the theoretical optimal MESW, that is, the MESW of the best routing algorithm applicable to the network, need to be determined to evaluate the optimality of a routing algorithm.

To determine to what extent packet energy P_e spent by a node n can be sustained by the power it harvests from the environment, a recovery time (T_e) is defined as the amount of time required by the energy scavenger to provide packet energy P_e to a node n. The recovery time is the ratio between P_e and the environmental power P_n available at the node n:

$$T_e = P_e/P_n \tag{7.9}$$

The workload can be considered energetically sustainable and the lifetime of the network is theoretically unlimited if the average interarrival time of packets on each edge is longer than the corresponding recovery time.

Recovery time directly correlates packet processing rate with available power. Channel capacity (C_e) of the maximum packet rate across an edge (e) is usually limited by the channel bandwidth, which represents energetic sustainability constraints. Hence, the channel capacity of edge e corresponds to its inverted recovery time:

$$C_e = 1/T_e \tag{7.10}$$

The flow F_e across the edge e is limited by its energetically sustainable channel capacity, which is expressed by

$$Fe \leq C_e = 1/T_e = P_n/P_e \tag{7.11}$$

Effective algorithms exist for determining the maximum flow between any pair of nodes in flow networks (networks with annotated channel capacities) by solving the so-called max flow problem [84]. It has been suggested that energy-harvesting WSNs can be viewed as flow networks and MESW problems cast into instances of max flow [85]. However, if Equation 7.11 is used to compute edge capacities, the overall maximum flow might end up being overestimated, due to the fact that all edges with the same source share the same power budget available at the common source node.

Therefore, a more general class of flow networks, called "node-constrained flow networks" [86], is needed to properly model the flow *Fe* across any edge *e* exiting from node *n*, which is limited not only by its maximum capacity, given by Equation 7.11, but also by the overall power budget of node *n*. This is expressed by

$$\sum_{e \text{ exiting from } n} F_e P_e \leq Pn \tag{7.12}$$

For only one outgoing edge node, Equation 7.12 is equivalent to Equation 7.11.

7.6 Summary

WMSNs have several factors that mainly influence their design, such as high bandwidth demand, integration with other wireless technologies, and power consumption. Prolonging the lifetime of operating battery and hardware power optimizations have been the focus of a vast amount of research in WMSNs. This research includes dynamic optimization of voltage and clock rate, wake-up procedures to keep high power-consuming electronics inactive most of the time, and energy-optimized protocol development for sensor communications.

Extracting energy from the environment is another important field. There are various forms of energy that can be harvested, such as thermal, mechanical, magnetic, solar, acoustic, wind, and wave. Additional sources of energy can help prolong the lifetime of sensor devices.

In this chapter, we looked into various ways of optimizing the power consumption of existing WMSNs. We also identified various unconventional sources of energy-harvesting based on available techniques and compared their advantages.

References

1. Gürses, E., Akan, O.B. Multimedia communication in wireless sensor networks. In: *Annales des Télécommunications* 60(7–8), pp. 872–900. Springer-Verlag, 2005.

2. Misra, S., Reisslein, M., Xue, G. A survey of multimedia streaming in wireless sensor networks. *Communications Surveys & Tutorials, IEEE* 10(4), 18–39, 2008.
3. Akyildiz, I.F., Melodia, T., Chowdhury, K.R. A survey on wireless multimedia sensor networks. *Computer Networks* 51(4), 921–960, 2007.
4. Aaron, A., Rane, S.D., Setton, E., Girod, B. Transform-domain Wyner-Ziv codec for video. In: *Electronic Imaging 2004*, pp. 520–528. International Society for Optics and Photonics, 2004.
5. Yang, Z., Liao, S., Cheng, W. Joint power control and rate adaptation in wireless sensor networks. *Ad Hoc Networks* 7(2), 401–410, 2009.
6. Chandrakasan, A.P., Brodersen, R.W. Minimizing power consumption in digital CMOS circuits. *Proceedings of the IEEE* 83(4), 498–523, 1995.
7. Gonzalez, R., Horowitz, M. Energy dissipation in general purpose microprocessors. *IEEE Journal of Solid-State Circuits* 31(9), 1277–1284, 1996.
8. Chang, J.H., Tassiulas, L. Energy conserving routing in wireless ad-hoc networks. In: *INFOCOM 2000. 19th Annual Joint Conference of the IEEE Computer and Communications Societies. Proceedings. IEEE*, vol. 1, pp. 22–31. IEEE, 2000.
9. Boice, J., Lu, X., Margi, C., Stanek, G., Zhang, G., Manduchi, R., Obraczka, K. Meerkats: A power-aware, self-managing wireless camera network for wide area monitoring. In: *Proc. Workshop on Distributed Smart Cameras*. 2006.
10. Atmel Corporation Datasheets. www.atmel.com/products/15.03.2009
11. Akyildiz, I.F., Su, W., Sankarasubramaniam, Y., Cayirci, E. Wireless sensor networks: A survey. *Computer Networks* 38(4), 393–422, 2002.
12. Miao, G., Himayat, N., Li, Y.G., Swami, A. Cross-layer optimization for energy-efficient wireless communications: A survey. *Wireless Communications and Mobile Computing* 9(4), 529–542, 2009.
13. Gupta, G., Younis, M. Load-balanced clustering of wireless sensor networks. In: *IEEE International Conference on Communications, 2003. ICC'03*, vol. 3, pp. 1848–1852. IEEE, 2003.
14. Polastre, J., Szewczyk, R., Culler, D. Telos: Enabling ultra-low power wireless research. In: *4th International Symposium on Information Processing in Sensor Networks, 2005. IPSN 2005.*, pp. 364–369. IEEE, 2005.
15. Alippi, C., Anastasi, G., Galperti, C., Mancini, F., Roveri, M. Adaptive sampling for energy conservation in wireless sensor networks for snow monitoring applications. In: *IEEE International Conference on Mobile Ad-Hoc and Sensor Systems, 2007. MASS 2007*, pp. 1–6. IEEE, 2007.
16. Raghunathan, V., Spanos, P., Srivastava, M.B. Adaptive power-fidelity in energy-aware wireless embedded systems. In: *Real-Time Systems Symposium, 2001.(RTSS 2001). Proceedings of 22nd IEEE*, pp. 106–115. IEEE, 2001.
17. Schott, B., Bajura, M., Czarnaski, J., Flidr, J., Tho, T., Wang, L. A modular power-aware microsensor with >1000x dynamic power range. In: *Proceedings of the 4th International Symposium on Information Processing in Sensor Networks*, p. 66. IEEE Press, 2005.

18. Alaei, M., Barceló, J. Power management in sensing subsystem of wireless multimedia sensor networks. In: *Wireless communications and networks: Recent advances*. Rijeka, Croatia: InTech—Open Access Company, 2012, pp. 549–570.
19. Rahimi, M., Baer, R., Iroezi, O.I., Garcia, J.C., Warrior, J., Estrin, D., Srivastava, M. Cyclops: In situ image sensing and interpretation in wireless sensor networks. In: *Proceedings of the 3rd International Conference on Embedded Networked Sensor Systems*, pp. 192–204. ACM, 2005.
20. Rowe, A., Rosenberg, C., Nourbakhsh, I. A low cost embedded color vision system. In: *IEEE/RSJ International Conference on Intelligent Robots and Systems, 2002*, vol. 1, pp. 208–213. IEEE, 2002.
21. Margi, C.B., Manduchi, R., Obraczka, K. Energy consumption tradeoffs in visual sensor networks. In: *Proceedings of 24th Brazilian Symposium on Computer Networks*. 2006.
22. Soro, S., Heinzelman, W. A survey of visual sensor networks. *Advances in Multimedia*, pp. 1–22, 2009.
23. Pistoia, G. *Battery operated devices and systems: From portable electronics to industrial products*. Elsevier, 2008.
24. Stojcev, M.K., Kosanovic, M.R., Golubovic, L.R. Power management and energy harvesting techniques for wireless sensor nodes. In: *9th International Conference on Telecommunication in Modern Satellite, Cable, and Broadcasting Services, 2009. TELSIKS'09*, pp. 65–72. IEEE, 2009.
25. Eliasson, J. Low-power design methodologies for embedded Internet systems. *PhD Thesis*, Department of Computer Science and Electrical Engineering, Luleå University of Technology, Luleå, Sweden, 2008.
26. Crompton, Thomas, PJ. *Battery reference book*. Newnes, 2000.
27. Jones, C.E., Sivalingam, K.M., Agrawal, P., Chen, J.C. A survey of energy efficient network protocols for wireless networks. *Wireless Networks* 7(4), 343–358, 2001.
28. Negri, L., Barretta, D., Fornaciari, W. Application-level power management in pervasive computing systems: A case study. In: *Proceedings of the 1st Conference on Computing Frontiers*, pp. 78–88. ACM, 2004.
29. Tsaoussidis, V., Badr, H. TCP-probing: Towards an error control schema with energy and throughput performance gains. In: *Proceedings of 2000 International Conference on Network Protocols, 2000*, pp. 12–21. IEEE, 2000.
30. Zhang, C., Tsaoussidis, V. TCP-real: Improving real-time capabilities of TCP over heterogeneous networks. In: *Proceedings of the 11th International Workshop on Network and Operating Systems Support for Digital Audio and Video*, pp. 189–198. ACM, 2001.
31. Ee, C.T., Bajcsy, R. Congestion control and fairness for many-to-one routing in sensor networks. In: *Proceedings of the 2nd International Conference on Embedded Networked Sensor Systems*, pp. 148–161. ACM, 2004.
32. Yaghmaee, M.H., Adjeroh, D. A new priority based congestion control protocol for wireless multimedia sensor networks. In: *2008 International Symposium on a World of Wireless, Mobile and Multimedia Networks, 2008. WoWMoM 2008*, pp. 1–8. IEEE, 2008.

33. Hull, B., Jamieson, K., Balakrishnan, H. Mitigating congestion in wireless sensor networks. In: *Proceedings of the 2nd International Conference on Embedded Networked Sensor Systems*, pp. 134–147. ACM, 2004.
34. Wang, C., Li, B., Sohraby, K., Daneshmand, M., Hu, Y. Upstream congestion control in wireless sensor networks through cross-layer optimization. *IEEE Journal on Selected Areas in Communications* 25(4), 786–795, 2007.
35. Murthy, C.S.R., Manoj, B.S. *Ad hoc wireless networks: Architecture and protocols*. Prentice Hall Publishers, May 2004, ISBN 013147023X.
36. Karl, H. An overview of energy-efficiency techniques for mobile communication systems. Telecommunication Networks Group, Technical University Berlin, Berlin, Germany, Tech. Rep. TKN-03-017 (2003).
37. Ehsan, S., Hamdaoui, B. A survey on energy-efficient routing techniques with QoS assurances for wireless multimedia sensor networks. *Communications Surveys & Tutorials, IEEE* 14(2), 265–278, 2012.
38. Cerpa, A., Estrin, D. ASCENT: Adaptive self-configuring sensor networks topologies. *IEEE Transactions on Mobile Computing* 3(3), 272–285, 2004.
39. Chen, B., Jamieson, K., Balakrishnan, H., Morris, R. Span: An energy-efficient coordination algorithm for topology maintenance in ad hoc wireless networks. *Wireless Networks* 8(5), 481–494, 2002.
40. Pottie, G.J., Kaiser, W.J. Wireless integrated network sensors. *Communications of the ACM* 43(5), 51–58, 2000.
41. Wu, J., Dai, F., Gao, M., Stojmenovic, I. On calculating power-aware connected dominating sets for efficient routing in ad hoc wireless networks. *Journal of Communications and Networks* 4(1), 59–70, 2002.
42. Doumit, S.S., Agrawal, D.P. Self-organizing and energy-efficient network of sensors. In: *MILCOM 2002. Proceedings*, vol. 2, pp. 1245–1250. IEEE, 2002.
43. Rabaey, J.M., Ammer, M.J., da Silva Jr, J.L., Patel, D., Roundy, S. PicoRadio supports ad hoc ultra-low power wireless networking. *Computer* 33(7), 42–48, 2000.
44. Shih, E., Cho, S.H., Ickes, N., Min, R., Sinha, A., Wang, A., Chandrakasan, A. Physical layer driven protocol and algorithm design for energy-efficient wireless sensor networks. In: *Proceedings of the 7th Annual International Conference on Mobile Computing and Networking*, pp. 272–287. ACM, 2001.
45. Alghamdi, M.I. PARM: A power-aware message scheduling algorithm for real-time wireless networks. In: *11th IEEE International Conference on Computational Science and Engineering Workshops, 2008. CSEWORKSHOPS'08*, pp. 299–306. IEEE, 2008.
46. Mangione-Smith, W., Ghang, P.S. A low power medium access control protocol for portable multi-media systems. *3rd International Workshop on Mobile Multi Media Communications*, September 25–27, 1996.
47. Stemm, M. Measuring and reducing energy consumption of network interfaces in hand-held devices. *IEICE Transactions on Communications* 80(8), 1125–1131, 1997.

48. Jayashree, S., Manoj, B.S., Murthy, C. On using battery state for medium access control in ad hoc wireless networks. In: *Proceedings of the 10th Annual International Conference on Mobile Computing and Networking*, pp. 360–373. ACM, 2004.
49. Yu, Y., Krishnamachari, B., Prasanna, V.K. Issues in designing middleware for wireless sensor networks. *Network, IEEE* 18(1), 15–21, 2004.
50. Zheng, R., Hou, J.C., Sha, L. Asynchronous wakeup for ad hoc networks. In: *Proceedings of the 4th ACM International Symposium on Mobile Ad Hoc Networking & Computing*, pp. 35–45. ACM, 2003.
51. Van Dam, T., Langendoen, K. An adaptive energy-efficient MAC protocol for wireless sensor networks. In: *Proceedings of the 1st International Conference on Embedded Networked Sensor Systems*, pp. 171–180. ACM, 2003.
52. Hohlt, B., Doherty, L., Brewer, E. Flexible power scheduling for sensor networks. In: *Proceedings of the 3rd International Symposium on Information Processing in Sensor Networks*, pp. 205–214. ACM, 2004.
53. Singh, S., Raghavendra, C.S. PAMAS—power aware multi-access protocol with signalling for ad hoc networks. *ACM SIGCOMM Computer Communication Review* 28(3), 5–26, 1998.
54. Singh, S., Woo, M., Raghavendra, C.S. Power-aware routing in mobile ad hoc networks. In: *Proceedings of the 4th Annual ACM/IEEE International Conference on Mobile Computing and Networking*, pp. 181–190. ACM, 1998.
55. Xu, Y., Heidemann, J., Estrin, D. Adaptive energy-conserving routing for multihop ad hoc networks. In: *Research report 527, USC/Information Sciences Institute*. 2000.
56. Stemm, M., Gauthier, P., Harada, D., and Katz, R. Reducing Power Consumption of Network Interfaces in Hand-Held Devices, In: *Proceedings 3rd Intl. Workshop on Mobile Multimedia Communications*, Princeton, NJ, 1996.
57. Polastre, J., Hui, J., Levis, P., Zhao, J., Culler, D., Shenker, S., Stoica, I. A unifying link abstraction for wireless sensor networks. In: *Proceedings of the 3rd International Conference on Embedded Networked Sensor Systems*, pp. 76–89. ACM, 2005.
58. Jacome, M., Catthoor, F. Special issue on power-aware embedded computing. *ACM Transactions on Embedded Computing Systems (TECS)* 2(3), 251–254, 2003. ISSN: 1539-9087.
59. Brown, C. Endless energy is harvesting's promise. EE Times. February 27, 2006. http://www.powermanagementdesignline.com/showArticle.jhtml?articleID=181400884
60. Hac, A. *Wireless sensor network designs*. West Sussex, UK: John Wiley & Sons, 2003.
61. Havinga, P.J.M., Smit, G.J.M. Energy-efficient wireless networking for multimedia applications. *Wireless Communications and Mobile Computing* 1(2), 165–184, 2001.

62. Liu, D., Svensson, C. Power consumption estimation in CMOS VLSI chips. *IEEE Journal of Solid-State Circuits* 29(6), 663–670, 1994.
63. Ho, R., Mai, K.W., Horowitz, M.A. The future of wires. *Proceedings of the IEEE* 89(4), 490–504, 2001.
64. Sylvester, D., Keutzer, K. A global wiring paradigm for deep submicron design. *IEEE Transactions on Computer-Aided Design of Integrated Circuits and Systems* 19(2), 242–252, 2000.
65. Raghunathan, V., Srivastava, M.B., Gupta, R.K. A survey of techniques for energy efficient on-chip communication. In: *Proceedings of the 40th annual Design Automation Conference*, pp. 900–905. ACM, 2003.
66. Lettieri, P., Srivastava, M.B. A QoS-aware, energy-efficient wireless node architecture. In: *1999 IEEE International Workshop on Mobile Multimedia Communications, 1999 (MoMuC'99)*, pp. 252–261. IEEE, 1999.
67. Chang, J., Ravi, S., Raghunathan, A. FLEXBAR: A crossbar switching fabric with improved performance and utilization. In: *Proceedings of the IEEE 2002 Custom Integrated Circuits Conference, 2002*, pp. 405–408. IEEE, 2002.
68. Chang, H., Cooke, L., Hunt, M., Martin, G., McNelly, A., Todd, L. Surviving the SoC revolution. *A Guide to Platform-Based Design*. Kluwer (1999).
69. Wang, H.S., Peh, L.S., Malik, S. A power model for routers: Modeling Alpha 21364 and InfiniBand routers. In: *Proceedings of the 10th Symposium on High Performance Interconnects, 2002*, pp. 21–27. IEEE, 2002.
70. Ye, T.T., De Micheli, G., Benini, L. Analysis of power consumption on switch fabrics in network routers. In: *Proceedings of the 39th Annual Design Automation Conference*, pp. 524–529. ACM, 2002.
71. Wang, H.S., Zhu, X., Peh, L.S., Malik, S. Orion: A power-performance simulator for interconnection networks. In: *Proceedings of the 35th Annual IEEE/ACM International Symposium on Microarchitecture, 2002. (MICRO-35)*, pp. 294–305. IEEE, 2002.
72. Dutta, P.K., Culler, D.E. System software techniques for low-power operation in wireless sensor networks. In: *Proceedings of the 2005 IEEE/ACM International Conference on Computer-Aided Design*, pp. 925–932. IEEE Computer Society, 2005.
73. Kijewski-Correa, T., Haenggi, M., Antsaklis, P. Wireless sensor networks for structural health monitoring: A multi-scale approach. In: *ASCE Structures 2006 Congress*. 2006.
74. Singh, A., Budzik, D., Chen, W., Batalin, M.A., Stealey, M., Borgstrom, H., Kaiser, W.J. Multiscale sensing: A new paradigm for actuated sensing of high frequency dynamic phenomena. In: *International Conference on Intelligent Robots and Systems, 2006 IEEE/RSJ*, pp. 328–335. IEEE, 2006.
75. Deshpande, A., Guestrin, C., Madden, S.R., Hellerstein, J.M., Hong, W. Model-driven data acquisition in sensor networks. In: *Proceedings of the 30th International Conference on Very Large Data Bases - Volume 30*, pp. 588–599. VLDB Endowment, 2004.

76. Jeličić, V. Power management in wireless sensor networks with high-consuming sensors. *Qualifying Doctoral Examination*, University of Zagreb, 2011.
77. Bult, K., Burstein, A., Chang, D., Dong, M., Fielding, M., Kruglick, E., Ho, J. et al. Low power systems for wireless microsensors. In: *International Symposium on Low Power Electronics and Design, 1996*, pp. 17–21. IEEE, 1996.
78. Hac, A. *Wireless sensor network designs*. West Sussex, UK: John Wiley & Sons, 2003.
79. Kulkarni, P., Ganesan, D., Shenoy, P., Lu, Q. SensEye: A multi-tier camera sensor network. In: *Proceedings of the 13th Annual ACM International Conference on Multimedia*, pp. 229–238. ACM, 2005.
80. Feng, W.C., Kaiser, E., Feng, W.C., Le Baillif, M. Panoptes: Scalable low-power video sensor networking technologies. *ACM Transactions on Multimedia Computing, Communications, and Applications (TOMCCAP)* 1(2), 151–167, 2005.
81. Paradiso, J.A., Starner, T. Energy scavenging for mobile and wireless electronics. *Pervasive Computing, IEEE* 4(1), 18–27, 2005.
82. Jiang, X., Polastre, J., Culler, D. Perpetual environmentally powered sensor networks. In: *4th International Symposium on Information Processing in Sensor Networks, 2005. IPSN 2005*, pp. 463–468. IEEE, 2005.
83. Mitcheson, P.D., Green, T.C., Yeatman, E.M., Holmes, A.S. Architectures for vibration-driven micropower generators. *Journal of Microelectromechanical Systems*, 13(3), 429–440, 2004.
84. Ford, L.R., Fulkerson, D.R. *Flows in Networks*, Princeton University Press, 1962.
85. Lattanzi, E., Regini, E., Acquaviva, A., Bogliolo, A. Energetic sustainability of routing algorithms for energy-harvesting wireless sensor networks. *Computer Communications* 30(14), 2976–2986, 2007.
86. Bogliolo, A., Lattanzi, E., Acquaviva, A. Energetic sustainability of environmentally powered wireless sensor networks. In: *Proceedings of the 3rd ACM International Workshop on Performance Evaluation of Wireless Ad Hoc, Sensor and Ubiquitous Networks*, pp. 149–152. ACM, 2006.

Index

F

G

H

I

J

L

M

Q

Y

Z